The Mcclure Family

MATTHEW THOMPSON McCLURE, Sr.
1834———
(Taken at Seventy-six.)

ALEXANDER McCLURD

Presses of Frank A. Owen,
Petersburg, Virginia.
1914.

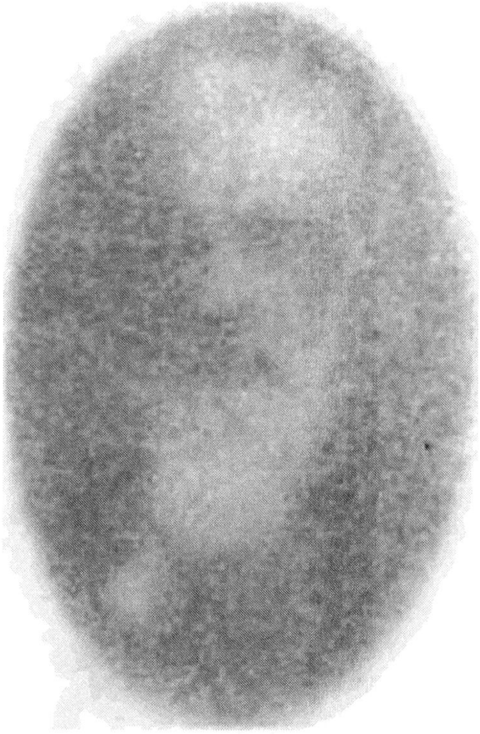

THE

MCCLURE FAMILY.

BY

JAMES ALEXANDER McCLURE.

LIMITED EDITION.

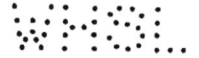

Presses of Frank A. Owen,
Petersburg, Virginia.
1914.

JOHN

FOREWORD.

THIS BOOK is an effort to preserve the names and something of the deeds of those who established the McClure family in America. While the result is far from satisfactory, I feel that I have rendered to the name in general, and to my own family in particular, a real service.

The work is the product of vacation days and rare leisure moments, thrown together rather than carefully arranged. It is the log cabin of our early ancestors rather than the modern mansion, to which I hope it will in time give place.

While all with whom it has been my privilege to converse or correspond have shown for the undertaking the greatest interest and concern, to whom I express my sincere appreciation, there are a number who have rendered special service and whose names I wish to mention in particular. First, the late Col. Charles McClure, of Ill., whose interest in the subject moved me primarily to the undertaking; Rev. A. D. McClure, D. D., Wilmington, N. C.; Prof. Geo. M. McClure, Danville, Ky.; Prof. C. F. W. McClure, Princeton University; Rev. James W. McClure, Cynthiana, Ky., Mr. Wallace M. McClure, Knoxville, Tenn.; Mr. Hugh S. McClure, New York City; Mr. Wm. A. McClure, Fairfield, Va.; Mrs. N. J. Baker, Nace, Va., Mr. Edward Frazer, Lexington, Ky.; Dr. J. D. McClure, London; Mr. John Wilfried McClure, Dublin.

The classification of the material which covers over two hundred years, seven generations, is as follows:

The first generation, born about 1700, is undesignated.

The second generation, born about 1733, is designated A, B, C, etc.

The third generation, born about 1767, is designated I, II, III, etc.

The fourth generation, born about 1800, is designated 1, 2, 3, etc.

FOREWORD.

The fifth generation, born about 1833, is designated (1), (2), (3), etc.

The sixth generation, born about 1867, is designated a, b, c, etc.

The seventh generation, born about 1900, is designated (a), (b), (c), etc.

There are doubtless errors and omissions other than typographical, to which readers will kindly call my attention.

It is my desire to have members from the various branches of the family send me from time to time all items of family interest, marriages, births and deaths, that they may be carefully filed as a basis of information for any future family record.

And may there be fulfilled unto us the prophecy of Jeremiah, who said unto the house of the Rechabites, "Thus saith the Lord of Hosts, the God of Israel, Because you have obeyed the commandment of Jonadab your father and kept all his precepts and done according to all that he hath commanded you, Therefore thus saith the Lord of Hosts, the God of Israel, Jonadab the son of Rechab shall not want a man to stand before me forever."

J. A. McCLURE.

Petersburg, Virginia,
October 15, 1914.

INTRODUCTION.

THE ORIGIN of the name McCLURE has been frequently discussed in the genealogical literature of Great Britain. The following theories have been advanced:

1. The name (variously spelt McClure, McCluer, McClewer, Maclure, McLewer, McLure, and McLuir), comes from the Gaelic word MacLobhair, pronounced MacLour, and means "son of the leper."

2. That it comes from the Gaelic MacGiolla-odhar (which in the genitive is uidhar and pronounced ure), contracted to MacIlure and hence McLure or McClure, and means "son of the pale one." This theory is advocated by Rev. Edmund McClure, M. A., London, Secretary of the Society for the Promotion of Christian Knowledge.

3. That it is a derivation of the Gaelic MacLeabhair, (pronounced MacLour) and means "son of the book," i. e. they were the teachers in the Clan McLeod, just as the McMcRimmons (or McCrimmons) were the clan pipers. They were MacLeabhair McLeods, McLeabhair (McLour, McLure, McClure), eventually becoming the sir-name. Several Gaelic scholars deny this derivation of the name, tho' admitting the very ancient tradition of the McClure tutorship in the Clan McLeod.

4 That the name is identical with MacLir (or MacLur) the seagod of Ireland and the Isle of Man. This theory is advanced in an article published in the Dublin University Magazine on the late Sir Robert McClure, the navigator.

5. The McClures were originally a Manx family, the first legendary king of the Island being a Manannan McClure, is the tradition inherited by the McClures of Manchester, England, to which family belong the late Sir John W. McClure, M. P., and the Very Rev. Edward C. McClure, D. D., Dean of Manchester. Held also by Sir Edward Stanley McClure.

INTRODUCTION.

6. That the name means "great bruiser." An ancient king of Scotland was attacked by highwaymen. One of his attendants so distinguished himself by his prowess that he was called MacClure, "Mac" signifying "great" as well as "son of." A blow from the fist is still known in Scotland as a clure.

7. That it originated in the ancient sport of Falconry, in which the lure was used to recall the falcon. The crest of this family of McLures is a hand in armour holding a falconer's lure.

8. A soldier from the ancient town of Lure in Normandy crossed over with William the Conquerer. He was rewarded for his service by a grant of land in the Island of Skye and was known as DeLure, Mac being later substituted for De, to harmonize with the Gaelic custom.

9. The theory advocated by Rev. J. Campbell McClure, Minister of Marykirk, Kincardinshire, Scotland, is that the McClures are a sept of the Clan McLeod. In addition to extant records in Galloway of the McClure family, showing it to be of McLeod origin, Mr. McClure states that the family tradition handed down to him through a long line of long-lived ancestors is, "In early times a sept of the MacLeods left the Island of Skye for Ulster, where the northern Irish slurred the 'd' of MacLuide (as it was then pronounced) into 'r,' hence, MacLure. Later many of the name passed over from the northeast of Ireland to Galloway, thus to Wigtonshire and so on to Ayrshire. These districts to-day contain many McClures."

It is certain that McClures are in some way connected with the Clan McLeod, evidenced by the fact that the oldest traditions of the family in Scotland take them back to the Isle of Skye; the traditions of Skye link together the McClures and the McLeods; McClures have always had the same motto, crest and tartan as the McLeods, and their right to them has never been called in question.

McClure history, then, properly begins with the McLeods.

INTRODUCTION.

Some authorities aver that they are of Irish descent. In an old volume of the Ulster Journal of Archaeology there is given a long pedigree of the McLeods, deducing them from various Scottish chieftains and princes, back to one Fergus Mor MacEarcha. The generally accepted theory is that they descended from Leod, one of the three sons of Olave the Black, King of Man and the Isles, tho' it is said that there is no documentary evidence extant to prove this claim.

Leod, born early in the 13th century, married the daughter of MacRailt Armuinn, a Norwegian chieftain, and by her acquired large possessions in Skye, including the fortress of Dunvegan, which is still in the possession of the family.

They held mainland estates under the Crown as early as 1340, and island estates at the same time under the Lords of the Isles. When the final forfeiture of the Lords of the Isles took place at the end of the 15th century, the McLeods got charters of their island estates from the crown.

Their name is conspicuous in Scottish history. They occupied the post of honor at the battle of Harlaw, 1411; they were at the battle of the Bloody Bay, 1485; they took part in the negotiations to transfer the allegiance of the Highland chiefs from the Scottish to the English king and signed the commission under which these negotiations were carried on. They took part in the battle of Worcester, 1661, led by Sir Norman McLeod of Bernera, where they lost 700 men.

Treated by Charles II with the grossest ingratitude, they took no part in subsequent Stuart uprisings, tho' there is a letter extant from James II, dated Dublin, 1690, imploring McLeod to join Dundee.

It is said that the name (Maklure) occurs in Scotland as early as the 12th century. A very old record is one of 1485, where Ewin MakLure and Gilbert MakLure witnessed a contract between Thomas Kennedy of Blaresguhan and Margaret Kessox, of Little Dunrod, Kirkcudbright. These McClures are supposed to have been friends (or relatives)

INTRODUCTION.

of the Kennedy (Seneschal) of Carrick in Scotland and cadets of the Carrick family of McLures of Bennane. These are all Galloway folk.

In the Acta Dom. Audit., published by the government in 1839, there is, under date of October 6, 1488, a decree that Johne Lord Kennydy, Johne of Montgomery and Michiell McLure shall devoid, &c., the lands of Barbeth to Janete Hamiltown. There is a record of January 24, 1489, that Johne Lord Kennydy, Johne of Montgomery, and Michell McClure shall pay to Janete Hamiltown, &c. (Note the two spellings of the name in this short extract.)

Barbeth is close to Kirkintulloch, northeast of Glasgow.

It is claimed by some that the original home of the McClures in Scotland was in the southwest, probably in Galloway.

Andrew McClure, late of Glasgow, now of London, states that Ayrshire is full of McClures. In Munsey's Magazine, February, 1911, in an article on Robert Burns, illustrated with a photograph of old Alloway Kirkyard, the name George McClure appears on one of the stones. Many of the family are buried here. Rev. J. Campbell McClure, Kincardineshire, Scotland, belongs to this family. There is a family tradition that one of his ancestors, an ecclesiastical reformer, suffered persecution under Charles II in those well known days when the heroic and faithful Covenanters were subjected to such unholy treatment. His home in Dalmellington was invaded and all his furniture taken out and burned.

A member of one of the Scottish families states: "The earliest ancestor we actually know of is Martin McClure, who lived at Balmaghil in Kirkcudbrightshire about 1750, where, I believe, he is buried. He had five sons: William, John, David, Robert and Andrew, all of whom came south, we being descendents of the eldest, and I know more or less of the descendents of the others. The crest and arms of our branch are,—

Arms: Argent, on a chevron engrailed azure, in chief

two roses, and in base a quatrefoil, gules, a martlet between two escallopes of the first.

Crest: An eagle's head, erased, proper."

Mr. Robert W. McClure, of the firm of Black, McClure & McDonald, Glasgow, writes, February 17, 1913: "My grandfather, James McClure, was a sea captain and I rather think was born in Bargany Estate, a few miles from Girvan. My father, James McClure, was Parochial Schoolmaster in Riccarton, near Kilmarnock, Ayrshire.."

Ian Maclaren has given a master's touch and added interest to the name in Scotland in his portrayal of William McLure, in A Doctor of the Old School.—"A tall, gaunt, loosely made man, without an ounce of superfluous flesh on his body, his face burned a dark brick color by constant exposure to the weather, red hair and beard turning grey, honest blue eyes that look you ever in the face, huge hands, with wrist-bones like the shank of a ham, and a voice that hurled his salutations across two fields, he suggested the moor rather than the drawing room. But what a clever hand it was in an operation, and what a kindly voice it was in the humble room when the shepherd's wife was weeping by her man's bedside. * * * * * He was 'ill pitten the gither' to begin with, but many of his physical defects were the penalty of his work, and endeared him to the glen. He could not swing himself into the saddle without making two attempts and holding Jess's mane, neither can you 'warstle' through the peat-bogs and snow-drifts for forty years without a touch of rheumatism. But they were honorable scars, and for such risks of life men get the Victoria Cross in other fields. McLure got nothing but the secret affection of the Glen, which knew that none had ever done so much for it as this ungainly, twisted, battered figure, and I have seen a Drumtochty face soften at the sight of McLure limping to his horse. 'His father was here afore him,' Mrs. McFadyen used to explain, 'atween them, they've hed the country side for well on tae a century.'"

INTRODUCTION.

>"Aye, dear Maclure; him maist o' a'
>We lo'e, and thro' the drifts o' sna',
>Unmindfu' o' the north wind raw
> We tearfu' come;
>Wi' a' the mourning glen to draw
> Near-haun his tomb.
>
>An' barin' there oor heids we pray
>That we may so live ilka day
>That when we come to pass away
> Frae a' things here,
>Truth may the tribute to us pay
> O' love wrung tear."

The scene of the doctor at the home of Tammas and Annie Mitchell is of peculiar interest to the McClures of Augusta county, Virginia, when it is remembered that the Mitchells and McClures were friends and among the first settlers of Augusta county.

The name is frequently found in Scotland to-day, and as in America, they are usually among the substantial members of their communities. Dr. John Watson, on his last visit to America, introductory to an address in Philadelphia, speaking of the Scotch families in the United States and their noble ancestry, mentioned especially the McClures and requested any of the name to come forward and speak to him at the conclusion of his address.

The late Earl of Stair, Scotland, states that the McClure family is one of the oldest in the list of the Scottish Untitled Aristocracy.

McCLURES IN IRELAND.

When and why did members of the family emigrate from Scotland to Ireland? There are two answers to this question.

First, *in the Planting of Ulster.* Following the accession of James VI of Scotland to the throne of England, 1603, the Earls of Tyrone and Tyrconnell inaugurated a general rebellion against the King. The effort failed, resulting ultimately in 511,465 acres of land, six counties in the province of Ulster, being forfeited to the crown.

James sought to settle upon these lands a Protestant population. Grants of land and numerous privileges were held out as inducements. Thousands availed themselves of the advantageous offer, and settled with their families upon these forfeited estates.

Among these emigrants were three McClures from Ayrshire, Scotland, supposedly brothers, who crossed over the channel to Ireland in 1608.

One settled at Saintfield, County Down. In this branch of the family the name Anthony frequently occurs, as it does also in an Ayrshire family of McClures "represented now as sole survivor by Mr. William McClure, Solicitor, of Hill Crest, Wigton, Scotland."

From him descended the late William Waugh-McClure, Justice of the Peace, Windsor Terrace, Lurgan, Ireland; also Thomas McClure born 1716, married in 1750 Elizabeth Ralston, the ancestor of John Wilfrid McClure, the author of several articles on The McClure Family published in the Belfast Witness, 1904, and from which the facts here stated are taken. He was connected for a number of years with the Munster and Leinster Bank, Dublin. This couple, Thomas and Elizabeth McClure, were weavers, the latter being famously deft in the use of the distaff. They died aged 101 and 102, respectively.

The most distinguished of the McClures of Down was Rev. Robert McClure, for sixty-three years pastor of the Presbyterian Church at Annahilt, ordained and installed

April 29, 1760. His great grandson, Prof. John Robinson Leebody, M. A., D. Sc., of Magee College, Londonderry, furnishes the following:

"My great-grandfather, the Rev. Robert McClure, was minister of Annahilt from 1760 to 1823. His family resided near Belfast where they owned some property. From this Mr. McClure derived income sufficient to live in easy circumstances. His staff of servants included a butler—not a usual luxury for a Presbyterian minister either then or now. His wife was a daughter of Archdeacon Benson, of Hillsborough, and a grand-daugher of a former Bishop of Down and Connor. Mr. McClure was on terms of intimacy with the country gentry and a great favorite with the Marquis of Downshire, with whom he used to dine every Wednesday at the castle. Many offers of promotion were made to him if he would consent to join the Episcopal Church, which he resolutely declined.

He had a numerous family, but of the history of several of them I have no details.

One of his daughters married Rev. Dr Wright, his assistant and successor. Another, John Robinson, a gentleman farmer near Hillsborough, who was my grandfather and after whom I am named. Another, Rev. Mr. Ashe, an English rector. One of his younger sons, Arthur, was in the army, and I believe attached to the staff of the Duke of Kent. After retiring from the army he resided near Lisburn, and I recollect being at his house when a boy. He used to tell that he had frequently carried our late Queen in his arms when she was a child.

Mr. McClure was the Moderator of the Synod 1779; preached from Philip., 4: 1-5. At the opening of the Synod in 1780 his sermon on I Timothy, 4: 16, was printed.

Mr. McClure is described as a man of distinguished appearance. I have heard my mother say that she remembered seeing him frequently when she was a girl, and her recollection of him was a tall, white-haired old gentleman with an ear trumpet. I may mention an anecdote I have heard which illustrates, amongst other things, the difference between the present standard of ministerial propriety and that current a century ago. Near Hillsborough, at a place called The Maze, there is a race-course, once very famous, and I believe still of considerable repute. Riding to it one morning during the race week with some of the local gentry, he overheard an altercation between a woman and her son, whom she was endeavoring to persuade to stay away from the races. To the final declaration of the youth, '*I will go; there is our minister going,*' Mr. McClure merely remarked, 'No one will ever say that again,' turned his horse round and despite the exhortations of his friends for being so needlessly scrupulous, rode home and was never seen on a race-course again."

The second brother settled at Crumlin, County Antrim, ancestor of the late Sir Thomas McClure, (1806-1893), M. P. from Belmont, son of William McClure and Elizabeth Thomson and grandson of Thomas McClure and Anne Swan, of Summer Hill, County Antrim; whose remote ancestor was an officer under William III, at the Battle of the Boyne.

The third brother settled in Armagh, ancestor of James McClure, of Armagh "attained 1688 in the reign of James II, along with quite a number of other Protestant landowners in Ireland." He is referred to as "James McClure, gentleman."

Dr. David Miller, pastor of the First Presbyterian Church of Armagh, writes under date of January 28, 1913: "My records of the 18th century, baptisms and marriages, are very defective, covering only the period 1707-28. I can find only one entry with the name McClure; it is a marriage— John McClure and Margaret Martin, June 13, 1723. The name does not seem to have been common about Armagh, nor is it yet. In the records of the Synod of Ulster I notice that an Elder, James McClure, attended the Synod at Antrim in 1705. His minister's name was Archibald Maclane, who could not have been the minister of the Armagh church, for his name was John Hutcheson, but his congregation was in the Presbytery of Armagh.

The second distinct emigration of McClures to Ireland

was from 1661 to 1688. Under the last two Stuarts the acts of oppression in Scotland were so severe and so continued that to escape them many sought a refuge with their countrymen, who had colonized Ireland in peaceful times. Crossing the channel in open boats, exposed to the greatest danger, they reached the friendly shores of Ireland and found a hearty welcome and homes free for a little while from the oppression that made them exiles.

In Boswell's "Tour through the Highlands," he mentions under date of October 16, 1773, meeting at the house of The Mac Quarrie, "Chief of Ulva's Isle," a Capt. McClure, of Londonderry, master of the Bonnetta sailing vessel. He says: "Capt. McClure was of Scottish extraction, and properly a MacLeod, being descended from some of the MacLeods who went with Sir Norman MacLeod, of Bernesa, to the Battle of Worcester, and after the defeat of the Royalists, fled to Ireland, and to conceal themselves took a different name. He told me there were a great number of them about Londonderry, some of good property."

We have another native of Londonderry in the person of Captain Robert McClure, born about 1775, "an officer in the old 89th Foot and served abroad." He saved the life of a fellow-officer, General le Mesurier, a gentleman of considerable property and a native of Guernsey, who afterwards became guardian to his son. Capt. McClure, while stationed at Wexford with his regiment, married in 1807 Jane, d. of Archdeacon Elgee. His son, Sir Robert John Le Mesurier McClure (1807-1873) was the discoverer of the North-west Passage, for which he received a large grant of money, the thanks of Parliament and knighthood.

A distinguished soldier died recently in Dublin in the person of William McClure-Miller, formerly of Ochiltree, Ayrshire. On his retirement from the service, he was appointed Governor of H. M. Prison at Arbour Hill, Dublin. He served in a regiment of Lancers and passed through no less than twenty-eight engagements on the foreign field, some of them the most important and decisive in the history of recent times. Like those of his name in general, he was long-lived and his death, though he had passed the four score limit, was hastened by an accident. Paternally a McClure, he was obliged on succeeding to some property to assume the second name of Miller. He left a son and namesake who is one of the clerks in H. M. Prisons' Board, Dublin Castle. A long and description memoir appeared in the Dublin papers at the time of his decease.

There are records on a tombstone erected at Findrum, County Donegal by Andrew McClure, Surgeon, Royal Navy, to the memory of his ancestors whose remains lie deposited in the vault beneath, and who for upwards of two hundred years had resided at Findrum. This tomb has inscribed upon it the coat-of-arms, crest, and motto of the McClures.

About 1850 there was published an article on Surnames in Down and Antrim, at which time not a single parliamentary voter named McClure lived in Co. Down and only three in Co. Antrim. There were many of the name living in East Donegal. "James" was a common name among them, but still more common was "Richard." There were

"Fiel Dick, Deel Dick and Dick of Maghesnoppin,
Red Dick, Black Dick and Dick who supped the broughin,"

all alive at the same time and all related.

Rev. W. T. Latimer, Eglish Manse, Dungannon, Ireland, writes June 6, 1913: "My great grandmother, Bell Kelso, died 1781, aged 58. A sister of hers, probably younger, married a Donegal McClure, whose Christian name I don't know. The family all went to America, including a daughter who married a Mr. Elliott. A sister, Susanna McClure, in 1764, married John Dill of Springfield, Co. Donegal, whose son Samuel was minister of Donoghmore, Co. Donegal. Captain McClure, the arctic explorer, belongs to this Donegal family."

In my judgment this Donegal McClure was Samuel who died in Rockbridge county 1779, leaving a wife, Mary, and among other children, Jean Elliott.

Mr. J. W. Kernohan, M. A., Secretary of the Presbyterian Historical Society of Ireland, writes on Aug. 5, 1913;

"We have not in our care church records of East Donegal, nor do I know of any except one in Magee College, Derry. I have searched the Muster Roll, 1631, and the Hearth Tax List, 1663, for Co. Derry, but did not find a single McClure. There is a register of Burt neighborhood (Co. Donegal, near Londonderry) for the years 1676-1719, but the searching of the book would mean some labor and time."

Rev. A. G. Lecky, author of "The Days of the Laggan Presbytery," writes August 7, 1913, of Co. Donegal:

"Among the names of the men who paid Hearth Tax in the Parish of Raphoe, 1665, there are two John McClures of Augheygalt, and a Gilbert McCluer in the adjoining Parish of Donoghmore. Also, amongst the names of Elders from Raphoe who attended meetings of the Laggan Presbytery between the years 1672-1700, are John, Arthur and Richard McClure. Also John McClure from Burt, near Londonderry. The name has always been a common one in this district. There are at present six McClure seat holders in the congregation of Convoy." Rev. Francis McClure of Carrigut, Co. Donegal, died in the United States some years ago while on a visit to his son.

Rev. John J. McClure, D. D., of Capetown, South Africa, writes September 9, 1913: "My father, Rev. Samuel McClure, who ministered at Crossroads, near Londonderry, and who died in 1874, came from Dernock, near Ballymoney, Co. Antrim, where his forefathers had been for some generations. They came originally from some place in the southwest of Scotland."

It is generally agreed that the Irish McClures are not one family, but are descended from a number of ancestors who emigrated from Scotland after 1608. They have not as a rule preserved their genealogies, hence the difficulty in tracing connection between them and determining the place and date of their origin in Scotland. The Ulster Journal of Archaeology, vol. II, p. 160, states that the center of the McClure families in Ulster is in Upper Marsareene, the most southern barony in Antrim. They are now all over Ulster, and those in one village or town are generally unconscious of any connection with those of another.

It is said of the descendants of the McClure who settled first at Knockbreda, near Belfast in Co. Down, 1608, that some of them settled in Belfast, some in Lisburn, Ballymena and other places in Co. Antrim. Some went further afield into Derry, which has several monuments of them. There are tombstones in the old burying ground at Knockbreda going back to the early years of the 18th century. In Carmany churchyard, Co. Antrim, there is the grave of Isabella McClure, daughter of Archibald McClure of Belfast, who died February, 1788, aged seven years.

At the Tercentenary Celebration of Presbyterianism in Ireland, held in Belfast June, 1913, J. W. Kernohan, M. A., in his address on Irish Presbyterianism, said in conclusion: "Indeed, in the commercial and professional annals of Belfast, it would be found on inquiry that if the names of the outstanding Presbyterians were eliminated, * * * it would be robbed of much of its moral, material and intellectual strength." In the list, which he gives in this paragraph, we find the name McClure.

Several of the Ireland McClures claim a coat of arms as do those of Scotland, generally similar to that of the late Sir Thomas McClure, of Belfast, namely, a domed tower and pennant, but while his motto was Spectemur agendo, theirs is Paratus sum, which is also the motto of the McClures of Lancashire, though their coat of arms and crest are different.

In addition to the arms, crests, &c., of the McLeods, which belong equally to the McClures, we find in Robson's Heraldry, Vol. II, and Fairbairn's Crests, Vol. I, under McLure, (or MacLure,) Scotland:

ARMS—Argent on a cheveron engrailed azure between three roses, gules, a martlet of the field.

CREST—An eagle's head, erased, proper.

MOTTO—Paratus sum.

INTRODUCTION.

Also,

ARMS—Argent, a cheveron, azure, between two roses, gu. in chief, and a sword, point downward, in base of the second.

Also,

ARMS—Argent, a dexter hand erased, fesseways gules holding a dagger, point down, azure, in chief three crescents, sable.

CREST—A domed tower, on top a flag, all proper.

MOTTO—Manu forti.

See also Burke's Landed Gentry.

The minutes of the Presbyteries and Synod of Ulster give some interesting information.

We find in the minutes of the Laggan Presbytery 1672-1700, the name of

Richard McClure, Elder from Donoughmore, 1679.
Arthur McClure, Elder from Raphoe, now Convoy.
John McClure, Elder from Raphoe, now Convoy.
Richard McClure, Elder from Raphoe, now Convoy, 1693.
John McClure, Elder from Burt, near Derry, 1698.

The name occurs frequently in the records of the General Synod of Ulster, 1691-1820.

James McClure, Armagh Presbytery, 1705, 12, 15.
John McClure, Belfast Presbytery, 1723.
Thomas McClure, Templepatrick Presbytery, 1733.
Daniel McLewer, Templepatrick Presbytery, 1738.
William McLewer, Templepatrick Presbytery, 1740.

McCLURES LEAVE IRELAND FOR AMERICA.

The period of Covenanter persecution in Scotland was one of comparative quiet in Ireland, but persecution came again under James II. And later in the reign of Anne (1702-1714), under the Test Act, they were made exceedingly uncomfortable. Unless they conformed in worship they could hold no public office, nor be married by their own ministers, nor bury their dead by their own simple rites, nor build churches, nor buy land, nor employ teachers except those of the Established Faith. Thus deprived by oppressive laws of every position of trust or honor, denied the liberty of speech, the free exercise of conscience, together with burdensome restraints on their commerce and extortionate rents from their landlords, they began to look toward America as another and a better home.

Says Froude: "In two years which followed the Antrim Evictions, 30,000 Protestants left Ulster for a land where there was no legal robbery, and where those who sowed the seed could reap the harvest." The government, alarmed at this depletion, gave relief and checked the emigration for a while. But in 1728 it began anew, and from then to 1750, it is estimated that 12,000 came annually from Ulster to America.

Physically and morally, of all the people in the world, these Scotch Irish were the best suited by nature and by Providential training for building up a new country. Some of them were scholars, as Robert Alexander, a Master of Arts of Dublin University, who, in 1749, built on land now owned by Samuel Finley McClure, near Old Providence Church, Augusta County, Va., the log school-house, sowing the seeds of learning, of which Washington and Lee University is the ripening fruit.

Landing in Pennsylvania, some of them crossed the Alleghanies and settled the western part of the State. Another stream flowed southward, entering the beautiful Shenandoah Valley, spreading over Augusta and Botetourt and

Rockbridge counties; and then some of them vainly dreaming that there could be upon this continent a more beautiful or fertile country, pushed on to Southwest Virginia, to Tennessee, Kentucky and farther west. Others turning eastward crossed the Blue Ridge and found homes in Southside Virginia, or pressing on, settled the piedmont section of the Carolinas.

That they were pioneers in the American Revolution and the struggle for religious liberty is an oft told tale.

Among the first of these settlers in the upper Valley of Virginia were a number of McClures. In writing of them and of their descendants, I fully agree with Sir Walter Scott, that "Family tradition and genealogical history are the very reverse of amber; which, itself a valuable substance, usually includes flies, straws and other trifles; whereas these studies being in themselves very insignificant and trifling, do, nevertheless, seem to perpetuate a great deal of what is rare and valuable in ancient manners, and to record many curious and minute facts, which could have been preserved and conveyed through no other medium."

And also the saying of Edmund Burke, "People who never look backward to their ancestors will never look forward to posterity."

THE McCLURE HOMESTEAD, AUGUSTA CO., VA.

Deeds

From George II to William Beverley, 1736.
From William Beverley to Sarah Ramsey, 1739.
From Sarah Ramsey to John Fulton, 1753.
From John Fulton to James Fulton, 1789.
From James Fulton to John McClure, 1819.
From John McClure to Matthew Thompson McClure, Sr., 1873.

McClures in Virginia.

JAMES McCLURE, the founder of the family in Augusta county, was born in the north of Ireland about 1690, came to America with his wife, Agnes, and five children, and settled in Long Meadow on Middle River of the Shenandoah, about five miles north of Fishersville. The first mention of his name is found in Hume's Old Field Book, page 53, "survey for James McClure, corner to Jno. Hart, in Geo. Robinson's line, 8 br. ye 18, 1738." His deed for land is recorded in Deed Book 3, p. 247, Orange Court-house, Va., and is as follows:

THIS INDENTURE, made the fifth day of June, in the year of our Lord one thousand and seven hundred and thirty-nine,

BETWEEN William Beverly, Gent. of the County of Essex, of one part, and James McLure, of the County of Orange, of the other part, WITNESSETH That the said William Beverly, for and in consideration of the sum of five shillings, current money of Virginia, to him in hand paid by the said James McLure at or before the Sealing & Delivery of these Presents, the receipt whereof is hereby Acknowledged, HATH Granted, Bargained & Sold, & by these Presents doth Grant, Bargain & Sell unto the said James McLure all that tract of Land Containing four hundred and eight Acres, more or less, being part of Beverly Mannor, BEGINNING at a white oak in George Robinson's line * * * * to the Beginning, AND all Houses, Buildings, Orchards, Ways, Waters, Water Courses, Profits, Comodities, Hereditaments and appertenances whatsoever, to the said premises hereby Granted, or any part thereof Belonging, or in any wise appertaining: And the Reversion & Reversions, Remainder & Remainders, Rents, Issues & Profits thereof, To HAVE AND TO HOLD the said Tract of Land and all & singular other the Premises hereby Granted, with the appertenances, unto the said James McLure, his Ex'rs, Adm'rs & Assigns, from the day before the date hereof, for and during the full Term & Time of one whole year from thence next Ensueing, fully to be Compleat & Ended: Yielding and paying therefor the Rent of one Ear of Indian Corn, on LADY DAY next, if the same shall be Lawfully demanded; to the Intent & Purpose, that by Virtue of these presents, and of the Statute for Transferring Uses into Possessions, the said James McLure may be in Actual Possession of the Premises, and be thereby enabled to accept & take a Grant & Release of the Reversion & Inheritance thereof to him & his heirs. IN WITNESS Whereof, the said William Beverly hath hereunto set his hand & seal the day & year first above written.

W. BEVERLY. (Seal.)

Sealed & Delivered in the presence of
James Porteus,
Thos. Wood,
John Latham.

On page 218, Order Book, 1739-'41, Orange Court-house, we find the following record of his importation:

At a Court held for Orange County on Thursday, the 24th day of July, 1740, James McClure made oath that he imported himself, Agnes, John, Andrew, Ellionor, Jean & James McClure Jun'r, at his own Charge from Ireland to Philadelphia & from thence into this Colony, and that this is ye first time of his proving his and their rights in order to obtain Land wc'h is ordered to be certified.

Hon. Joseph A. Waddell, LL. D., in an address at the celebration of the 150th Anniversary of Augusta Stone Church, October 18th, 1899, said of these first settlers: "They came on foot or horseback and could bring little with them besides tools and implements of labor, and seed corn obtained in older settlements in Pennsylvania. Each family located according to its will and pleasure, not troubling themselves about land titles, and after erecting rude cabins, set to work to clear and cultivate the land. For at least a year the first comers must have subsisted on wild meat, the deer and other game which abounded, without bread or any substitute for it. During the first twelve or fifteen years the dwellings were hardly better furnished than the wigwams of the Indians. There were no tables, chairs, knives and forks, glass or chinaware, and many things now found in the humblest homes. The mention of 'cart wheels and tire' in an inventory of 1746 is the first intimation of a wheel-vehicle in the settlement. But horses and cattle were numerous and "the big ha' Bible, was found in nearly every cabin."

Of the life of this pioneer we have but little information. He was a charter member of Tinkling Spring Presbyterian Church, organized 1740, as shown by its sessional records. His name is signed to the following:

"Know all men by these presents yt we ye under Subscribers Do appoint and Constitute our trusty and welbeloved friends, Colonel Jas. Patton, John Finley, George Hutchison, John Christian and Alex'r Brackenridge, to manage our publick affairs, to Chuse & purchase a piece of ground to build our meeting house upon it, to collect our minister's salary and to pay of all Charges Relating to said Affairs, to get pay of the people in proportion of this & to replace seats in our said meeting house, wch we do hereby promise to Reimburse them; they allways giving us a months warning by an advertisement on ye meeting house Dore and a majority of the above five persons provided all be apprised of theire meetting; theire acting Shall Stand these persons above named Shall be accountable to ye minister and Session twice Every yeare for all theire proceedings Relating to the whole affair to which we Subscribe our names in the presence of the Re'd Mr. Jno. Craig.

August ye 14th, 1741, Copia Vera."

The Baptismal Register, 1740-1750, kept by the pastor, Rev. John Craig, and now in the possession of Gen. John

E. Roller, Harrisonburg, Va., gives the record of the baptism of his two youngest children, viz:

"Samuel, child of James McClure, baptised Nov. 7, 1740.
Esther, child of James McClure, baptized Nov. 8, 1741."

The Chalkley Records, Vol. II, p. 28, show that there was a schoolhouse on James McClure's land, built 1747, "at the foot of the hill in the meadow." So far as I have any information, this is the oldest record of a schoolhouse in Augusta county.

In the record of the court proceedings of Augusta county, 1749, he appears as plaintiff, with a minute to the effect that he and the defendant having agreed out of court, the case was discontinued.

On p. 222, Order Book 2, 1751 he and his wife, Agnes, appear in court as witnesses for John Finley, and received for their services 150 pounds of tobacco each.

His will is recorded in Book 3, p. 47, Augusta County, and is as follows:

"In the name of God. Amen. The twenty third day of September, 1756, I, James McClure of South Carolina, Tylor, being very sick and weak in body but perfect mind and memory thanks be given unto God therefor calling unto mind the mortality of my body, and known that it is ordained for all men once to die, do make and ordain this my last will and Testament, that is to say principally and first of all I give and recommend my soul into the hands of Almighty God that gave it and my body I Recommend to the earth to be buried in Decent Christian burial at the discretion of my Executors nothing doubting but at the general Resurrection I shall receive the same again by the Mighty Power of God and as touching such Worldly Estate wherewith it hath pleased God to bless me in this life, I give, demise, dispose of the same in the following Manner and form.

Imprimis, I will and ordain that my son James shall have my Bible and big pot.

Etem, I will and ordain that my son Samuel have the next bigest pot, I will and ordain that my wife Agness have the use of both pots during her life. I will and ordain that my Movable Estate shall be equally divided between my dearly beloved wife Agness and son James and my son Samuel. I will and ordain that my plantation Shanado, be equally divided between my two sons, James and Samuel. I will and ordain that my son James pay to my son John one shilling Sterling. I will that my son James pay to my son Andrew one shilling Sterling. I will that my son James pay to my daughter Eleanor a shilling Sterling, and a shilling Sterling to my daughter Jean and a shilling Sterling to my daughter Esther, and I make and ordain Wm. Givens and William McClure my sole executors of this my will to take care and use the same perform in witness whereof I have hereunto set my hand and seal the day and year above written.

JAMES McCLURE, (L. S.)

Signed sealed & delivered by the said James McClure as and for his last will and testament.

William Hanna,
William Beard,
William McClure.

At a court held for Augusta County, August 18, 1761, this last will and Testament of James McClure dec' was proved by William Hanna, one of the witnesses thereto, who also made oath that he saw William Beard & Wm. McClure the other evidences sign the same, and that they are since dead and John McClure Eldest son and heir at law of the said deceased, having appeared in Court & declared he had no objection to proving the said will, it is admitted to record and Wm. McClure one of the executors being dead and William Givens the other Executor residing in South Carolina, Administration of with the will annexed, is granted James McClure son of the said Deceased he having with security complied with the law."

His giving his residence as, "South Carolina, Tylor," calls for explanation.

Probably the terrible drought that prevailed in the Valley during 1755, together with Gen. Braddock's defeat July 9th of the same year, leaving the settlers unprotected from the Indians, caused him to move temporarily to South Carolina where other friends and relatives had settled. "The consternation was universal, and many of the settlers on the western frontier fled across the Blue Ridge, and even to North Carolina."—Annals of Augusta County, p. 109.

"Tylor" probably should have been written, "tailor," indicating that in addition to his farming he followed a trade.

His two older sons who had families in Augusta County certainly did not accompany him. He doubtless returned to Augusta previous to his death, 1761. His grave is probably among the many unmarked around Tinkling Spring Church.

The following is from The Watchman of the South, June 7, 1844:

"The settlers of the great Valley of Virginia congregated in neighborhoods, and laid their dead side by side, near their places of worship, unless some hard necessity turned them to another place of sepulture. * * * * * You may find the resting place of the Pioneers of the Valley by following the footsteps of the immigration as it advanced from the Potomac to the Catawba. * * * * * Often, very often, in the midst of the forest of graves, you cannot tell where rest the men and women who had courage to lead the way in settling beyond the mountains, and give what Governor Gooch desired for his province, a line of defence against the savages, buying immunity in their religion, by the freedom from fears and alarms and massacres, their bravery conferred, at their own peril, on the settlers below the Blue Ridge.

In some few spots you are among the ancestors. Were the graves to give up their dead, and the dust be fashioned into bones and sinews and put on flesh like the forms that mouldered there, you would gaze on the determined visage of the men and the calm decision of the matrons—the toil-

worn frames, the labor-hardened hands of a generation that
loved a church without a prelate—a generation that fled
from the oppression that harrassed their ancestors for cen-
turies, and like them ungovernable in their demands for the
unalienable rights of man—a generation of men and women
that accomplished in their poverty what wealth cannot purchase, and reared another generation to hazard death for
freedom of conscience and liberty of person, "for a State
without a king." Such a one you may find at TINKLING
SPRING in Augusta County, Va.

Going down from the splendid prospect from Rockfish
Gap to the edge of the "late country," as the Sage of Mon-
ticello termed it, you enter the bounds of the oldest congregation in Augusta; one that contends with Opequon for
the honor of being the first in the great Valley, and the first
in the State after the days of Makemie—the congregation
of the Triple Forks of the Shanandoah, which formerly
stretched across the Valley from the Gap to the Western
Ridge in the horizon. You are, too, in the bounds of that
division of the congregation named Tinkling Spring, which
assembled to worship God in the Southeastern part of the
congregation, the Old Stone Church being the place for as-
semblage for the Northwestern part of the settlement tra-
versed by the paved road. Ministers were few and men
went far to worship; far as it would now be estimated, as
then eight or ten miles were an ordinary ride—or walk—for a
Sabbath morning to the house of God.

But we were searching for the graves of the settlers.
Come to this yard to the west of the church, surrounded
by a stone wall in the shape of a section of a horseshoe.
* * * * Come down now to this Southwest end. In the
irregular piece of ground, surrounded on three sides by a
stone wall, full of mounds, but not a single inscription.
Here is the resting place of the ashes of the ancestors of
many families in Virginia and Kentucky. Men whose
names are woven by their descendants in the web of political and religious courts in colors too vivid to be unnoticed
or mistaken. Here are the sepulchres of the men that

turned the wilderness into habitations, waiting for the coming of the Son of God when the graves shall give up their dead. Let no spade or mattock ever hereafter disturb the soil that vegetates so luxuriously over their ashes. The time is coming—is already at hand—when history shall present to the living the actions of these venerable dead, and posterity will glory in deriving their descent from the settlers of the Valley of Virginia."

LINE OF JOHN McCLURE.

James and Agnes McClure left seven children.

A. JOHN McCLURE, the oldest, was born in 1717, as shown in Chalkley, vol. II, p. 5; came to Augusta county with his father 1738, and settled on South River, near Lyndhurst, six miles south of Waynesboro. His farm joined Edward Hall and John Coulter. His deed for 359 acres, dated March 1, 1749, is recorded D. B. 2, p. 692, Staunton, Va.

His name appears frequently in the extant records of the county.

On the muster roll (spelled McClewer) 1742 of Capt. Jas. Cathreys's company, tenth in the list.

On August 21, 1757 he, with others, was directed to clear a road from Edward Hall's to Wm. Long's mill.

He was a juror 1767.

On December 15, 1778, he deeded to Andrew Alexander, a son-in-law, 204 acres "being part of the plantation John McClure now lives on, on the waters of the South river cornering John Coulter and Edward Hall."

The last mention of his name in the county record is August 28, 1791, when he appeared as a witness in a case of Edward Hall vs. John Coulter. He was then 74 years of age. He died intestate about 1798. The date is fixed by a court record of June 19, 1798, when George Hutchison, his son-in-law, was appointed guardian of his youngest child, an invalid daughter, Eleanor.

The family records do not give the name of his wife, but

from known facts it is practically certain she was Elizabeth Steel, daughter of Andrew Steel, who died in Augusta county 1764.

The family Bible now owned by M. T. McClure, Sr., Spottswood, Va., gives the name and date of birth of their eleven children:

I. ANNE McCLURE, born October 27, 1745.
Baptized by Rev. John Craig, Nov. 10, 1745.
She married Geo. Hutcheson. Daughter,
1. Margaret, b. Feb. 19, 1785, d. Oct. 27, 1870.
She mar. Nov. 10, 1807, Isaac Hutchinson.
Issue.
(1). George Hutchinson, father of the late Henry Hutchinson, of Staunton, Va.
(2). John Lewis Hutchinson, of Point Pleasant, W. Va.

II. ESTHER McCLURE, b. Aug. 6, 1747, baptized by Rev. John Craig, Sept. 13, 1747, died Sept. 18, 1747.

III. JAMES McCLURE, b. Sept. 24, 1748, baptized Nov. 6, 1748, and died in Augusta Co. 1784. See Chalkley, Vol. I, p. 236.

IV. JEAN McCLURE, b. Jan. 6, 1751, d. single 1837.

V. ELIZABETH McCLURE, b. Aug. 1st, 1753. She m. Dec. 29, 1790, Francis Alexander. Among her descendants was the late Frank Alexander of Staunton, Va.

VI. MARTHA McCLURE, b. July 31, 1756. She married about 1778, Andrew Alexander, Jr., (son of Andrew and Catherine Alexander).
Issue.
1. Catherine Alexander, who m. Jan. 1, 1798, Jas. Arbuckle, of Greenbrier Co.
2. John Alexander, whose descendants now live on, or near, the farm where John McClure settled.

VII. MARY McCLURE, b. Nov. 14, 1758. She is possibly the Mary (Polly) McClure, who m. Thomas McCullough April 5, 1808. Chalkley, Vol. I, p. 396, gives a suit, John McClure vs. Thomas McCullough. "Writ 10th Jan., 1788. About 1783 defendant removed to the French Broads." He was living in Blount county, Tennessee 1795.

VIII. JOHN McCLURE, b. March 25, 1761, died April 2, 1761.

IX. MARGARET McCLURE, b. Nov. 21, 1764 and m. April 7, 1796, Rev. John McCue officiating, Andrew Henderson, son of Samuel and Jane Henderson, of Augusta County. They emigrated early in the 19th century to Blount County, Tenn. The writer of the following letter addressed to Mr. John McClure, Junior., Augusta County, Virginia State, Waynesborough, is probably her son.—

"State of Tennessee, Greene County, March 25, 1819.

Respected Counsin, I embrace this opportunity of writing to you, to inform you that I am in good health at present, trusting these lines will reach you and find you in the same. There has been great affliction in father's family, death has visited the family. Brother Samuel departed this life the 19th of this month, he was taken with the flux the first day of the month. Father is lying very low at this time, the flux appears abated, an old complaint appears to grow more fatal. The rest of the family is in tolerable health at present. Grandfather and Aunt Anne are in good health they both wish to be remembered to you and all the relations. Brother George and Aunt Anna has returned from Blount County. Mr. Whites and Uncle Hutchisons familys has been afflicted with the fever. I will mention the names of those that have been afflicted with the fever. Mrs. White, Sally, Anna Jorden, Eliza Drusilla, James Diddle. Uncle William was confined to bed six weeks, John White lay in the fever five weeks and continues very low. Miss Peggy White is in good health and also requests me to remember her to you. It appears that the journey has been a benefit to her and

Grandfather and Aunt Anna. I have heard that Mr. Fulton has sold their plantation. You may inform them that there is a plantation in Blount County for sale, five hundred and thirty-two acres of land nicely improved, one hundred of bottom land, all of good quality, the price six thousand dollars. A number more plantations in this County for sale. Please remember me to all enquiring friends, Miss Betsy Fulton in particular, please write the first opportunity the particulars of the place. I have sent two letters to Augusta, one to Uncle Daniel Henderson, another to Cousin John Alexander. I have not received any answer yet—another letter to William Thome. My reason for not writing to you sooner I was waiting to hear of Uncle William settling himself he has not purchased land yet. Aunt Anna wrote a letter to Miss Betsy Fulton, and has received no answer yet, we wrote the letters in January that we sent by the mail. Please to give respects to your father-in-laws family that is Mr. George Pilson. Grandfather and Aunt Anna wishes you to remember them to George Pilson and family. We have had a remarkable warm winter and it appears like being a very sickly season in this country. I wrote this letter in a hurry.

I add no more at present but remain your affectionate friend and well wisher.

ALEXANDER HENDERSON.

JOHN McCLURE, JR."

XI. ELEANOR McCLURE, b. September 15, 1769, Invalid, died single. Geo. Hutcheson appointed her guardian June 19, 1798.

X. ANDREW McCLURE, b. July 18, 1767, and died at the home of his son, John, near Old Providence church December 30, 1847. His grave is marked in Bethel Cemetery.

He married, on January 15, 1789, Mary Mitchel, fourth child of Thomas Mitchel and Elizabeth (McClanahan) Moore, Rev. Archibald Scott, pastor of Bethel church, officiating. His wife died 1795, and from that time until a few years before his death, he seems to have had no settled home, living with his sisters and children, frequently walking back and forth from Waynesborough to Old Providence, a distance of twenty-five miles, even when past seventy years of age. Unlike his father and his sons, he owned but little property. In personal appearance, he is said to have been a large, muscular man strickingly like his son John, whose photograph appears in this book.

The only extant reference to him is in a letter to his son John, from Samuel Coursey, dated, Xenia, Ohio, August 11, 1817. "How is your old father? I was going to write to him, but he talked of going away from our house and I did not no where and I thought he would not get it, but tell him I am wide awake and ask him, if he wins as much tobacco off Wm. Hutchison and the rest of the fellows about there as he used to do."

M. T. McClure, his grandson now living (1914) near Spootswood, Va., remembers him very distinctly, and states that while he sometimes imbided too freely he was a constant reader of the Scriptures. He also remembers hearing his grandfather and father talk of the battle of Point Pleasant, and is distinctly of the opinion that his grandfather told him, though only fourteen years old, he was present and participated in the battle of Guilford Courthouse. We know that Colonel George Moffett early in 1781, led a battalion of Augusta county men to North Carolina and participated in this battle. However I find no record of his ever having applied for a pension.

He was a soldier of the War of 1812, as shown by the following documents now in the hands of the writer, also the records of the Departments at Washington:

"Know all men by these presents that I, Andrew McClure, late a soldier in Capt. Thomas Sangster's Company in the Twelfth regiment of Infantry, who was enlisted the first day of March, 1814, to serve during the war and honorably discharged from the army of the United States March 30, 1815, as will more fully appear by my original discharge hereto annexed, have and by these presents, do nominate, constitute and appoint my son, John McClure, my true and lawful attorney for me and in my name and in my behalf to procure and receive from such officer, person or persons as shall be lawfully authorized to grant the same, a WARRANT for the quantity of land to which I am entitled for the service rendered by me as a private soldier in the army of the United States during the late war," &c.

In testimony whereof, I have hereunto set my hand and seal at Staunton, in the county of Augusta and State of Virginia this first day of August, 1816.

ANDREW McCLURE. (Seal).

Signed, sealed and delivered in presence of John Bumgardner, John Alexander and Wm. Clarke."

The War Department furnishes the following: "Andrew McClure enlisted March 4, 1814, at Staunton, Va., as a private of Capt. Sangter's Company, 12th U. S. Infantry, and was discharged March 30, 1815, at Fort Covington, by reason of the expiration of his term of service."

The Department of the Interior, General Land Office, gives the following: "On March 4th, 1819, military bounty land warrant No. 20883 for 160 acres was issued to Andrew McClure, private in Capt. Sangster's Company, 12th U. S. Infantry, war 1812. The warrant was located Nov. 27, 1820. Patent was issued to the soldier Nov. 27, 1820, and is recorded in Vol. 5, p. 291, Patent Records."

There is a tax receipt of February 2, 1825, of ten dollars on this property. Other than this it seems that the owner never took possession of the land.

Andrew McClure and Mary Mitchel had five children.

1. JAMES McCLURE, b. Augusta county, Va., November 4, 1789, and died at his home near Rogersville, Tenn., April 11, 1866. He emigrated about 1815 to Ashe county, N. C., removing a year or two later to Lebanon, Russell county, Virginia, thence to Tennessee. The following letters give in brief outline this period of his life:

"JEFFERSONTON, N. CAROLINA, June 23, 1816.
DEAR BROTHER:

I embrace this preasent opportunity of informing you that I am well at preasant and all my family, which is But Small. My wife gives her best Respects to you. I am out of the way of hearing from you by way of private opportunity. I received a letter from you during this Spring, which was the only acct I have had for some time. I received a letter from Sister Polly in the month of May, which was the last acct I had from your Country for some time. I have no knews from this Country that is worth writing. I have some idea that I shall Take a ride to the Western Country the latter part of the summer or winter, and if I find that I can be pleased with the Country, I shall in all probability leave this Country as soon as possible. I do intend to erect a tan-yard between this and next Spring if I should leave this Country, which I think I will at this time. If I should go I want Brother Thomas to go with me when he is done learning his trade. I think he can do as well, and better perhaps, than whare he is. I wrote a letter to him some time ago. If I can get what money is owing to me I can establish myself some whare that perhaps I can make a living. If Thomas had about three or four hundred dollars in Cash with what I shall have, if he should be inclined to go with me, we could make a grand Establishment in the western Country. If Thomas should be inclined to go with me if you have any money that you will not want to make use of, if you will lend it to him, I will be his Security for any amount you can let him have, that is provided he should engage in any business with me. A mechanick can make but little by working Journey work, unless he can get the management of a tan-yard, which is not easy got. I do not wish to persuade Thomas to do anything that would not promote his interest. As he is not settled nor myself as yet, I should be glad to have him with me, as our Occupations will suit to go together, and perhaps we might by doing business, be an advantage to each other. I should be glad if we could all settle near each other; as none of us is settled yet, it would be a great satisfaction to me if it could be the case. I have enclosed a letter to our father, give it to him the first opertunity. I shall write to you this Summer some time and by that Time I shall be determined. Write me when you receive this letter. Give my respects to all my friends. No more at preasent.

JAMES McCLURE.

JOHN McCLURE, Waynesborough, Va."

"LEBANON, RUSSELL COUNTY, VA., August 13, 1820.

Brother, I send you a few lines to inform you that we are all well at present hoping that this may find you all in the same situation. I have nothing to write to you, Polly and myself arrived safe home in five days from the day we left Mateers, which was very hard riding. Polly stood riding much better than I expected. We rode 103 miles in two days. Thomas will move in three or four weeks. Times is very tough in way of money, property is selling very low. I find it very hard to sell leather for money, how I may do this fall and winter I cannot tell. I have a great deal of work to do this fall. Tell Betsy I will write to her shortly. Give my respects to Aunt Betsy and all the rest of the Family. Tell Betsy that Thomas Alderson and Mary Jane Hanson was married while I was in Augusta, at last. If you don't think proper to write to me occasionally you may let it alone.

JAMES McCLURE.

MR. JOHN McCLURE, Greenville, Virginia."

"LEBANON, RUSSELL COUNTY, VIRGINIA, Feb. 15, 1828.

Dear Brother, I now write you a few lines at this time, to let you know that we are all yet alive and in good Health at present, and shall be glad to hear that my letter may find you all enjoying the same Blessings. I heard from Thomas McClure two weeks ago, him and his family was well at that time, the last account I had from Polly she had a very severe spell of sickness, was then on the Recovery. Thomas is going to move from Scott the first of March to Hawkins County in Tennessee, as I have understood. I have not seen him for some time, I expect to go and see him before he moves. Tell Betsy I received her letter a few days since and will write to her before a great while. I have no knews of Consequence to write to you.

Thare was two murders committed in Scott County this winter. Money is very scarce in this Country. I find it very hard getting along, give my Respects to Mr. Fulton & Aunt Betsy and your wife and Family. No more at present. JAMES McCLURE.
MR. JOHN McCLURE, Augusta County, Virginia:
The post office at Greenville."

"LEBANON, RUSSELL COUNTY, VA., Sept., 10, 1830.

Dear Brother, I received your letter on yesterday informing me of your well fare. We are all in good Health at present hoping this may find you in the same state of health.

I expect to move from this place in ten or twelve days to Hawkins County, Tenn. You will hereafter write to me thare. Give my best respects to Aunt Betsy, tell her I will write to her after I move. Give my Respects to your wife and children, My Father, Uncle Thomas and Family. Write to me in the course of a month. I have nothing more to write at present.
JAMES McCLURE.
MR. JOHN McCLURE, Greenville, Va."

As shown by the above letters he settled in Hawkins county, Tennessee, in the fall of 1830, having bought a farm of 500 acres, 2½ miles west of Rogersville, bordering

JAMES McCLURE,
1789-1866.

the Knoxville and Bristol "stage" road. Here in a large log house he kept a tavern, and in addition to his farming operations ran a tan-yard. About 1847 he built a large brick house on the hill top, a noble monument for that part of the country in ante-bellum days.

The advent of the railroad put an end to stage coach travel and the need of a tavern. Giving all his attention to his farm he became, according to universal agreement, the best farmer in Hawkins county. He was especially noted for his excellent fences with painted gates, which he required all passers to use. His motto was "Good fences make good neighbors," and consequently keep his in excellent repair.

In personal appearance he was large and muscular, with the characteristic McClure voice that could be heard across his farm. In politics he was a Jackson Democrat; in religion both he and his wife were "blue stocking" Presbyterians, and tho' in his latter years he rarely ever attended services, having become totally deaf, he saw that his family went in the "carry all" to Rogersville every Sabbath.

His letters above are written in a neat, bold hand showing that while his spelling and grammar are not up to the present day standards, he excelled in penmanship, evidently having had considerable early training. He was a man of superior intelligence, well posted and a great reader.

He is remembered as a man of bold independence of action as well as of thought, for despite his deafness he always insisted upon walking in the middle of the road. It was this independence that caused his death, being run over and seriously injured by a company of Union Cavalry, from which he never fully recovered. He and his wife are buried in the family plot near his home.

He married November 28, 1815, Susan Montgomery, (May 17, 1791—March 22, 1876), of Washington County, Va., Rev. Edward Crawford, of the Augusta family, then pastor in Washington Co., officiating. She was possibly a granddaughter of William Montgomery, who emigrated from Augusta to Washington Co. in 1769. Tall and handsome in appearance, she was a fitting helpmeet to her strong and pioneering husband.

Her father, Richard Montgomery, a Revolutionary soldier, lived near Meadow View, Va., where he was an Elder in the Rock Spring church. Her mother was Elizabeth McCall.

James and Susan (Montgomery) McClure had ten children.

(1). Mary Anne and Eliza Jane, twins, born July 27, 1817, and died Jan. 1, 1826, and Dec. 16, 1817, respectively.

(3). Elizabeth, b. Apr. 21, 1819, d. May 12, 1877.

She m. about 1844, Alexander Mason Doak, born at Tusculum, Tenn., March 26, 1819, and died August 22, 1903. He was a son of Rev. Samuel W. Doak, founder of Tusculum College, and Sarah Houston McEwen, and grandson of Rev. Samuel Doak, D. D., founder of Washington College, Tenn., and Hester Montgomery, both originally of Augusta Co.

In the McClure burying ground near Rogersville, Tenn., there is a stone bearing the following inscription:

"In memory of our father and mother
Rev. A. M. Doak,
Aged 84 years.
Elizabeth McClure,
Aged 58 years.
"They died as they lived, Christians."

Eight children, viz:

a. James M., b. Apr. 14, 1845, and died in the Confederate Army 1863.

b. Sarah A.

The following obituary gives the outline of her life:

"Mrs. W. A. Kite, of Johnson City, Tenn., died on August 14, 1912, at the home of her brother-in-law, Mr. W. C. Wells, of Marvin, Tenn., and her body lies buried in the cemetery of that village.

She was born October 2, 1846, married Captain W. A. Kite, who survives her, at St. Clair, Tenn., November 2,

1870. For the last thirty years they have lived in Johnson City.

Mrs. Kite, who was Miss Sallie A. Doak, was a granddaughter of Rev. Samuel A. Doak, D. D., whose name is eminent in the religious and educational annals of Tennessee. Washington College and Tusculum College are both monuments of his pioneer and constructive zeal, and his works still follow him in the sterling, spiritual qualities of the people of this section. Mrs. Kite was true, in her life and character, to this fine strain in her blood.

She was converted at the age of thirteen, and was ever afterward a faithful and diligent member of the Presbyterian church. Her unselfish and intelligent fidelity and her fervent prayers were a source of encouragement and power to the people of God among whom her lot was cast. She is deeply mourned by a large number of devoted friends. She is survived by four sisters and two brothers.

Her patience in long and painful illness, and the confidence with which she awaited her end, were born of deep and well-grounded convictions in a heart that knew and loved Him to whom she had committed her all.

G. G."

- c. Sue V., b. April 14, 1848. Lives single at Greenville, Tenn.
- d. Samuel H., b. Jan. 12, 1850. Not married.
- e. Mary A., b. May 30, 1853, m. J. J. Morrison, lives Romeo, Tenn.
- f. Endora E., b. Dec. 14, 1854, m. W. C. Wells, lives Marvin, Tenn.
- g. Alice F., b. Feb. 1, 1857, m. F. A. R. McNutt, lives Festus, Mo.
- h. Robert H., b. Jan. 17, 1859, m. Cleopatra White, lives Johnston City, Tenn.

(4). Mitchell, b. in Russell Co., Va., Aug. 16, 1820, emigrated with his father to Hawkins Co., Tenn., where he died Jan. 22, 1876, a prominent farmer and a respected citizen. He was a Southern sympathiser, but was not a soldier of the Civil War.

He m. on Dec. 26, 1843, Beersheba Cobb Kyle, of a prominent Hawkins Co., family. Eleven children.

 a. Sara Alice, b. June 1, 1845, living in Knoxville, Tenn.

 b. Jos. K., b. April 22, 1847. Lives near Whitesburg, Tenn. A good farmer, a loyal Presbyterian and a staunch Democrat. He married, Oct. 13, 1870, at the bride's home near St. Clair, Tenn., Hilah Morrisett, Rev. John W. Bachman, D. D., now of Chattanooga, Tenn., officiating. His only daughter, Beersheba Alice, who lives with him, was b. Aug. 12, 1877. His son, Richard Hugh, was born Aug. 26, 1879, and died May 19, 1895.

 c. Beersheba Ann, b. April 7, 1849, d. Aug. 16, 1849.

 d. James A., b. Dec. 29, 1851, d. Nov. 10, 1855.

 e. Absolom K., b. July 16, 1854, now of Pennington Gap, Va., Farmer. He married, Oct. 13, 1887, at her mother's home in Hickory Flats, Va., Mattie W. Carnes, her brother, Rev. J. W. Carnes, officiating. They have no children.

 f. John B., b. Aug. 21, 1856, and d. Jan. 18, 1903. His widow lives in Memphis, Tenn.

 g. Mary A., b. Dec. 11, 1860, m. the late———Sevier, who d. in Knoxville, Tenn., 1905.

 h. William K., b. Nov. 9, 1863, m. June 18, 1889, in the Church Street M. E. Church, South, Knoxville, Tenn., Rev. J. W. Bachman, D. D., officiating, Eliza Parsons Lewis. She was b. Jan. 18, 1870. Her father S. Duff J. Lewis, son of a Methodist minister, was a soldier in the Confederate Army, and later connected with a wholesale house in Knoxville. Her mother, Helen Wallace Arthur, of Covington, Ky., was of a prominent and wealthy family, tracing her ancestry through her paternal grandmother to the Scotch nobility. Her brother was for a number of years a member of Congress.

Issue:

(a). Wallace Mitchell McClure, b. July 30, 1890. B. A. University of Tenn., 1910; Bachelor of Laws (U. of Tenn.,) '11; Student Harvard Law School 1911-12; Student University of Wisconsin 1912; Student Columbia University, candidate for Ph. D. 1913. Member Phi Gamma Delta Fraternity; President of Chi Delta Literary Society and a representative in debate against University of Cincinnati '07, and Texas, '11. Editor-in-Chief Tennessee University Magazine 1909-'10, Bennet Prize for Essay '11 and McClung Medal for Moot Court work '11.

(b). Margaret Duff McClure, b. Oct. 6, 1892. B. A. University of Tenn., 1913, Chi Omega fraternity and Phi Kappa Phi honorary fraternity.

(c). William Kyle McClure, Jr., b. Dec. 4, 1894. Student University of Tenn. Member of the Sigma Alpha Epsilon fraternity.

(d). Robert Lewis McClure, b. Nov. 12, 1909.

i. Margaret Rice, b. Aug. 9, 1865, living single in Knoxville.

j. Hugh Walker, b. Jan. 9, 1868, m., 1899. Travelling Salesman, Albany, Ga.

k. Andrew Fulton, b. July 16, 1870. He owns "Horseshoe Bend," the old home place near Rogersville, Tenn., where he is a successful farmer. He is a Presbyterian "now and for always." In politics, a Democrat.

He married, Oct. 14, 1903, in the Royal Oak Presbyterian Church, Marion, Va., Sallie Phipps Miller, daughter of Daniel C. Miller and Charlotte A. Phipps, the Rev. J. McD. A. Lacy, officiating.

They have two children (a) Charlotte Miller, born in Marion, Va., March 15, 1905, (b) Mary Fulton, b. "Horseshoe Bend," Rogersville, Tenn., Oct. 2, 1907.

(5). Nancy, b. June 16, 1823, visited when a girl, her uncle in Augusta Co., d. s. Jan. 27, 1882.

(6). Montgomery, b. March 25, 1825, and d. June 3, 1892. He was also a farmer in Hawkins Co., where he lived and died. Like his brother, Mitchell, he was a Southern sympathier, but was not in the service.

He mar., June 16, 1864, Cynthia A. Johnson (Oct. 20, 1841—Jan. 14, 1884). Both are buried in the family plot near Rogersville.

Issue:
- a. John Sandford, b. June 24, 1865, m., 1906, lives in New York City.
- b. Sarah Johnson, b. July 24, 1867.
- c. James Andrew, b. Sept. 30, 1869, married and lives in Knoxville, Tenn.
- d. Anna C., b., Nov. 13, 1872, d. June 16, 1904.
- e. Susan Louisa, b. Nov. 4, 1874, d. May 9, 1904.
- f. Thomas Mitchell, b. Nov. 4, 1878, was accidentally killed, Seattle, Washington, April 14, 1907.
- g. Mattie Lee, b. Sept. 27, 1883, m. R. R. Goodman and lives at Church Hill, Tenn.

(7). Mary A., b. May 8, 1827, m. ——— Dunlap, and died at Knoxville, Dec. 10, 1901. Son, George.

(8). Virginia, b. June 16, 1830, d. Oct. 19, 1877.

(9). Margaret, b. Mar. 13, 1832, and d. Sept. 10, 1904. She m. Oct. 13, 1850, Col. James White, a prominent citizen of Hawkins Co. The following letter to John McClure of Augusta Co., his wife's uncle, is of interest:

"MATAMORAS, MEXICO, Oct. 28th, 1868.

DEAR UNCLE:

You may be somewhat surprised to receive this communication from this place, but so it is. The President nominated, and on the 28 of July last I was confirmed by the United States Senate Consul at this port, and on arrival here on the 9th Sept. received my commission and exequatur and commenced the discharge of my official duties as a representative of the U. S. Government. I have had my ups and downs such as I suppose are common to all Americans among foreigners, but I do think in my particular case the obstructions on the Road to Jordan have been wonderfully augmented,—unable to speak the language of natives, I am not only cut off from conversational & social enjoyment, but stand like a deaf and dumb man when others are happy and full of mirth (perhaps a part of it at my expense). Add to this the heat of a climate almost insufferable, water warm and impure, musketoes of the most improved breed and unlimited quantity; fleas friendly and abundant, and Mexicans who can and will upon the shortest possible notice steal not only your every article of personal property, but I verily believe could nearly steal a chew of tobacco out of your mouth without your notice. I have had one complimentary visit from them in this way. I was one night robbed of every particle of clothing, watch, valice, boots, shoes, socks and even to my looking glass, wash bowl & pitcher. This was verry soon after I opened my office as consul and was intended I suppose as a polite compliment to myself and government, and to show me how appreciative of my presence the natives were. In all they stold from me about $400, leaving me like Adam found himself in the garden—naked.

I am much disappointed in the character and inducements to stay in the place & will at the earlyest day I can do so resign my position & go home to my dear Mag and children, where I can at least be happy if I am not making so much.

I have no local or general news that would be of any interest to you escept it be that War and Revolution seem to be the order of the day here, which is always the case in this Republic; the citizens heare are looking for revolution now every day, but of this I of course, fear nothing. I have Uncle Sam's Stars & Stripes waving over my office. An emblem of nationality everywhere; a protection, & everywhere respected.

Well, now I have nearly consumed my sheet and have written almost nothing. How is Aunt, how is Consin Sally, her husband, her two sweet little children; how is Tom, his clever little rebel wife and baby? In a word, how are you all, is a question I would love to have answered.

I wish you to especially remember me to each and every one of these, but above all, to Aunt and Consin Sally; tell Tom his baby must be named Andrew Johnson. Or *See Some More* (Seymour).

When you have an opportunity to do so remember me to Andrew & familey in Staunton, and especially to Phebe; poor child, for a long while after I left Andrew's I could scarce fail to think of her & her affliction. My kindest regards and compliments to Mr. Bumgardner & family. All of whom were verry kind to me when in Greenville. I will, I think, be at home in December and probably go to Washington. Soon after which, if I do, I will try to again give myself the pleasure of a visit to your hospitable roof.

With love to all and assurance of my highest respect and esteem, I have the honor to be verry truly your obdt. servt. JAMES WHITE, U. S. Consul.
Matamoras, Mexico.

Mr. John McClure.

There were several children. Among them:
- a. James White, Jr., who m. a Miss Lincoln of Marion, Va., now lives in Oklahoma,—the parents of Lincoln and James White, (3rd) Mattie Lee, a star graduate of the Mary Baldwin Seminary, Staunton, Va., now married and living in Walla Walla, Wash., Margaret and others.
- b. Ida White, who now lives near Rogersville at the old White homestead, which is the site of the first house built by James McClure on settling in Tenn.

(10). Martha, b. Sept. 4, 1836, d. Sept. 22, 1901. She corresponded as long as she lived with her first cousin, M. T. McClure, of Spottswood, Va.

2. MARY McCLURE, b. in Augusta county 1791, emigrated to Russell County, Va., 1820, where she m. John M. Hendricks, of the same family as Vice-President Thomas A. Hendricks. Later moved to Calaway county, Mo. The following letters were written to her brother, John McClure.

"ESTERVILLE, SCOTT CO., VA., June 23, 1823.
BROTHER.

I take this opportunity of writing you a few lines to let you know that we are well at present. I have nothing particularly to write to you. I was at a great camp meeting last week in Lee. It was the greatest meeting that ever I was at; you might hear them shout a mile. There is no other society here but Methodist. I like them better than any other society. There were some fine meetings here at the courthouse.

Tom calls his son James Alexander. James has built him a fine house; have not been to see him for a year. I intend going there in a few days. I like Scott better than Russell. The people here are more moralise. Times is tolerable hard here at this time. I look for James here in a few days. Tom is doing tolerable good business here. Write to me how many children you have, what you call them all. No more at present. When I write to you again I have more to write.

POLLY McCLURE.

JOHN McCLURE."

"HAWKINS COUNTY, TENN., June 7, 1830.
DEAR SIRE.

I drop you a few lines to let you now that we ar in good health at this time, hoping thes lines may find you and your family enjoying your health. I have nothing particular to write to you at this time. I want you to tell Mathew Pilson that twenty dollars of that money that he let me have proved to be counterfit. A twenty dollar note on the United States Bank payable at Charlestown. I want you if you pleas to tell him to write to me amediately what I must do about the note. I want him to send me twenty dollars of good money, as I have lately bought me a tract of land and am very nedey for the money. Tell him not to fail in writing to me as soon as possible. And while you ar telling him be shoure not to forget to write yourself and let us now how you ar all doing. Polly will write to

Aunt Betsy in a few days concerning Betsy's affairs according to her request. So we remain yours with respect,

JOHN M. HENDRICKS & POLLY HENDRICKS.

To JOHN McCLURE."

HAWKINS COUNTY, TENN., June 15, 1833.

BROTHER.

I take my pen to let you know that we are well; hope thes few lines may find you the same. John went to see the new country this spring. He saw Tom in Indiana; he has nothing, everything he has is sold. He is a distrest man; he has not got his children. It would make you sorry to see the distress of Tom. James is going to move here this fall to this place. Want you to write to me when you get this letter. No more.

POLLY HENDRIX.

MR. JOHN McCLURE."

"CALAWAY COUNTY, Mo., April 20.

DEAR BROTHER.

I sit down this day to let you now that we are all well at this time and hoping that these few lines may find you all well at this time. I have met with a great loss by fire. I had everything in the world burned but my houses; nearly every pannel of fence I had was burnt and the two adjoining farms were also burnt. There has been a great destruction here by water; the Missouri river was so high that it was from ten to fifteen feet deep over all the bottoms. It swept off all the houses on the river. A great many lost everything they had. Corn and wheat crops failed here last season. The reason was it was so wet that people could not tend their crops; as to my part I did not rase hardly any. I have my meat and bread both to buy this year. I think I would of had enough to do me if I had not got it burnt.

Dear brother I have a great deal of trouble and difficulty in getting along. There is a great deal of sickness and death here. There is a disease here that is something like the collery. There is not a person that takes it that gets over it; no person lives over twenty-four hours. I have got in twelve acres of corn this year. Dear brother, I would be very glad and thankful if you would send me ten dollars in a letter. I don't think there would be any difficulty in sending it in a letter. I would not have called on you for any assistance, but I have nothing to buy meat and bread with; it can't be got for nothing but the money. I wrote to Brother James to sent me ten dollars. I met with the great loss in March. I send my best respects to your family and my father and Aunt Fulton. Write to me as soon as you get this letter.

MARY HENDRICKS.

John McClure."

"Callaway County, Mo., June 26.

Dear Brother.

I sit down this evening to let you now that we are all well at the preasent time and hoping that these few lines may find you all enjoying good health. I received your letter this morning and fifteen dollars in it. I am very glad and thankful to you for it, for I stood in great need of it. I have in twelve acres of corn this year; corn and wheat crops are very good here this season. Corn and meat is very scarse here. My neighbors were good and kind enough to help me to fence in twelve acres. I got the money you sent by Mr. Stuart, that, and what Brother James sent me saved my land. I have seven children living, four sons and three daughters. My oldest daughter is dead. My neighbors say they will help me to fence in my farm. I have three head of horses and three milk cows. I think after I get my land fenced in I can make a support. My oldest boy is fifteen years old. My boys allways made me a support until last year; the crops failed here last year.

I like to live in Missouri very well, but there is a great deal of sickness here; nearly every person has the chills and fever here in the fall. I live about ten miles from Uncle Bily Mateer; his children are all married but two. I would be glad to see you in this country. Brother James was out here about three years ago. Tobacco is six dollars a hundred here. This is a great country for to make sugar. I made about 200 pounds of sugar here this spring. Everything is cheap here but corn and bacon. I send my best respects to Ant Betsy and my father and to you and your family. I want you to write whenever you can for I am glad to hear from you all. I intend to put some of my boys out to a good trade after this year. I got a letter from Brother James in March. No more at present, but remember your sister,

MARY HENDRICKS.

JOHN McCLURE."

These letters were written about 1845.

In a letter from Mr. Daniel Nolley to Mr. Andrew Stuart, Waynesborough, Va., dated Fulton, Mo., March 30, 1841, I read, "you mentioned in your letter to the Major that Mrs. Hendrick's brother would like to know her situation. I was much pleased to see that they think of her and that there is some probability that they can render some assistance. Mr. Hendricks had when he died a verry pretty tract of land worth about $1,000; of available means to the estate, exclusive of land there is but little —say $100. I cannot tell you how much the estate owes, but I suppose it would amount to about $700, depending on a settlement yet to be made with a man who was his partner in a corn speculation. It will therefore require about $600 to save the land of Mrs. Hendricks which, with the property she has, will enable her to live comfortably. I feel much interest on her account, and if her friends can help her with that amount, it will do more for the comfort of a helpless family than could be done in almost any other case. Will you go and see them in reference to this matter and get them, if they can do anything for her, to do so as early as convenient. Will you write me on the subject as soon as they can determine this matter ?"

In a letter to same party from Mr. James Steele Henderson, dated Fulton, Mo., 8th March, 1847, I read, "P. S. Mrs. Hendricks, sister of our friend John McClure, you know, resides near here. Her husband died in debt and left her in rather destitute condition with large family of children. Her brother in Tennessee has helped her some. I would be gratified if you could see John McClure and ask a small friendly aid. I feel much interest in her behalf.

H."

I regret after several efforts I have not been able to communicate with any member of this family.

Of her four sons and three daughters I have no definite information. Mr. John N. McCue, of Auxvasse, Mo., states that Mr. Andrew Hendricks, of Bachelor, Missouri, probably belongs to this family.

3. ELIZABETH McCLANATHAN McCLURE b. in Augusta Co., 1792, and died single in Hawkins Co., Tenn., 1829, where she was keeping house for her brother Thomas after the death of his first wife, Phoebe Hendricks.

The following letters written by her to her brother John in Augusta Co., are of interest not only from the standpoint of family history, but for the insight they give into the conditions and ideals of her generation.

"SCOTT COUNTY, VA., March 28, 1828.
BROTHER AND SISTER:

I now take my pen to inform you that I am well at present and hoping that these few lines may find you all enjoying the same blessings. We got landed here last Tuesday about 1 o'clock, I was very tired travelling, I thought it a very long road. We had very good weather all the time, but Sunday morning it rained on us a little bit, Thomas family is all well, but the young man that lived with him took the fever in a week after he left home and is very sick at this time, the doctor has give him out since we came home but he appears to be better this morning, it does make very much against him he is lying at his hous wher he boarded, they wood pay no attention to him, they wood let him ly there days and not make his bed, you may judge from that what sort of people they is here. Thomas went down to his place yesterday to see about his things, the man was not able to go when he was gone, he will be home this evening. He cant move till this man gets so that he can leave him. This is a very disagreeable place to me I am glad Thomas is going to move from here. I haint got acquainted with many yet some of thes Ishmalites come in to see this sik man, they run in like they ware scared and never speak. We have a rocken chair I just sit and rock and look at them. They wont wait for a introduction—is that your sister? Yes sir—it is funny to hear them. I want you to rite to me when you get this letter how you all is—how Ant Betsy is and all the children is and how Sally is and how old Ant Moor is, I long to hear from you all. We haint had no word from James since we came home. Give howdy to Ant Betsy and uncle and all our friends and to your father. Rite and direct your letter to Rogersville we wont be here for a letter to come. I will write more particulars the next time. No more at present but remain your sister,

BETSY McCLURE.

JOHN AND JANE McCLURE."

"HAWKINS COUNTY, TENN., NOV., 2, 1828.
BROTHER AND SISTER:

I have taken up my pen this morning to let you know that I am still in the land of the living and in good health, thanks be to the great giver of all mercies for it. I haint much news to rite to you at present, Thomas and family is well, he has had a great deal of trouble about his business he is not improving his tanyard fast. We had a very dry summer since harvest. Corn is better than we expected, fall grain was very good, wheat is fifty cents a gushel in trade, corn won dollar a barrel. Money is very scarce in this Country. I have had a great deal to do for some time back. Thomas has had the workmen at his house. I have had no body to help till last week, we hired a black woman for a while. Thomas was up to see James about too weeks ago they wir all well then, he and his wife had the fever, his wife was very bad he had but a slight turn, his blak family had it and one died. Thomas had his daughter up there going to school all summer, he brought her home with him. Tell Pheby and Peggy Mitchell that I received their letters evening before last. You rote to me that you had plenty of fine flax dont forget me and save some till I come home, flax is not plenty here they work all on cotton here and that dont soot me. This is Court week here and we have some boarders, it does keep me busy for I have always a pack of men to work for. Some times I wish there was no men. Pheby says that the children is learning very fast I am glad to hear it, don't forget Sally she is my favorite child among all. Tell Sis to learn to right fast and rite to me, take my love to yourself. No more but remain your effectionate sister until death, BETSY M. McCLURE.

JOHN AND JANE McCLURE.

DEAR ANT:
I rite you a few lines to let you know that I have not forgot you yet. I have had my health very well since I left home, I got very lean in the summer but I am like the rabits begining to faten when the white frost comes. I was sorry to here of Ant Moores death and of so many deaths of the sore throat. Thomas treats me very well, he gets me anything I want, he thinks he never would have got along if I had not been with him. I have done a great deal for him it has been a great charge on me, but I hope the Lord will enable me to walk in the Christian path to fulfil my duties to them, it has caused me to have some serious times. My trust is strong in the Lord that he will guide me in all difficulty. Send me word when you want me to come home. Please to except my love for I must conclude and bid you good evening. No more but remain your neise until death.

BETSY M. McCLURE.
BETSY FULTON."

"HAWKINS COUNTY, TENN., January, 27," 1829.
DEAR BROTHER AND SISTER:
I wonst more take up my pen to let you know that we are all well. I have understood you is all getting well

again and I was very glad to hear it. Brother Thomas is away from home at this time, he is gone with horses to Carolina, he had some of his own and he bought some more and got them low and is to pay part of them in trade, they wir very nice saliable horses and in good order. He started about ten days ago, I dont look for him for six weeks, I have all his business to attend. I wish you and Jane would come out and stay all night with me and let me see your fine son whether he is worth ownen. John, you and Jane is doing good business when you have two sons, no wonder you raised a hundred bushels of potatoes. I suppose if you all live to another year you will double your measure and old Mary another bit of a girl. Mr. and Mrs. Torbet and family is her this night. Give howdy to Uncle Thomas and family and write to me what has become of Mary Mitchell, I have heard of all my friends but her. Tell Mathew and Mary that I received thir letters. I want to know what is become of Billy Moor. I have understood that he has rented out his place, if I had been at home would have tried to put that notion out of his head. Want to see you all very bad and the children. Jane, I send as much callico as makes Mary and Sarah bonets and George gallowses, if I had nown that Torbet was going to Augusta I would have had some caps for Ant and you. Give howdy to all the Children for me and take my love to yourselves. No more. Your sister.

BETSY McCLURE.

JOHN AND JANE McCLURE."

UNCLE AND ANT:

I rite a few lines to let you know that I have not forgot you yet, I feel satisfied to here that you is spared in the land of the living, for I have herd of so many old people died since I left home and young as well as old. I have great reason to be thankful, so far I have my health very well all fall and winter, feel that I cant return thanks enough for it. I feel a poor unworthy creature, many such kind mercies bestowed upon me, but I still trust and hope that the great giver of all these kind mercies will guide

JOHN McCLURE,
1794-1873.
JANE PILSON,
1797-1882.

us all in that path of duty if we will only use the means that is sent to us but when we do our best it is too little. I think Thomas trys to fulfil his command in religion, he keeps worship twist a day. I think he is a good Christian in heart but him and me cant agree in sentiments, I do not like the ways of their church, I dont get much to preachen. I just think sometimes that if I was at Bethel and here Mr. McFarland it would revive my cold spirit. I must conclude and bid you good evening. No more your nies.

BETSY McCLURE.

JAMES AND BETSY FULTON."

She died suddenly about a year after writing the above letter and is buried in the family lot near Rogersville, Tenn.

4. JOHN McCLURE, b. near Waynesboro, Va., May 28, 1794, and died at his home near Old Providence Church April 26, 1873. His grave is marked in the Bethel Cemetery.

His mother, dying when he was less than two years old, he was raised by a great aunt, Sarah Steele, of Augusta Co.

He m. July 27, 1819, Jane Pilson (June 14, 1797—Sept. 18, 1882), dau. of George Pilson and Elizabeth Thompson, removed the same year to the home of his uncle and aunt, James and Betsy Fulton, which he later inherited and where he lived for fifty-six years, one of the most prosperous farmers and highly respected citizens of that section of the county.

His father being practically without property, his early educational advantages were limited. He had, however, some school privileges, as did his brothers and sisters. His English Grammar published at Holgate, near York, 1795, is in the hands of the writer.

He was twice elected Ruling Elder in Bethel, but his conception of the office was such that he could never bring himself to feel that he was worthy to accept it.

In 1829, in company with Mr. John B. Christian and his brother-in-law, Mathew Pilson, he made a trip on horseback to Russell and Washington Counties, Va., then through Kentucky, Ohio and Indiana, a distance of about 1800 miles. The following letter was written to his wife while on the trip:

"SULLIVAN COUNTY, INDIANA, Oct. 30, 1829.
MY DEAR AND AFFECTIONATE WIFE.

I just take holt of my pen to inform you that I am well at preasent and hoping that these few lines may find you and the rest of the family enjoying the same like blessings. We are now at William McCutchan's, all enjoying good health. I was very much disappointed in not getting a letter at London; I was verry anxious to hear from you when I was there, as I expected a letter. We are going to start to-morrow morning on strait to the Missouri. Our horses has stood it tolerable well. My horse got gravel, but is better. Mr. Christian's horse has a very sore back. It is a little uncertain whether we will be home agin Christmas. The roads are verry bad traveling, there has been a great dele of raine. I have seen a heap of fine country. I saw Nathaniel Steele at Robert McCutchan's in Ohio. They were all well; I was glad to see them. Wm. McCutchan's family are all well.

If ever we live to get home and see you I can tell you a good deal, and hope it is the Almity's will that we will see each other again in this world. No more, but remain your affectn husband,

JOHN McCLURE.

JAIN McCLURE."

I give below a number of letters found among his papers. In addition to the bits of family history they contain, they are interesting in giving something of the conditions of a hundred years ago. The following are from Capt. Samuel Steele:

"2d June, 1817, GREEN BRIER, VA.
DEAR SIR.

I want you to send the axes and the publick rifles; make a twist of straw and rop all along and send moles

and wipers and Fanny's saddle, and Fanny wants all the thread that Betty has except the blanket yarn; all the wool and flax and cottin thread & wants Sally to keep her spinning wool if she has nothing else to spin. Fanny wants as much of the wool sold as will get as much cambrick as will make a spread and as much musling as will line it. If you can, get it five quarters wide. I want you to keep old Tom in the meadow, as he ain't worth anything in the field. We are well. Send the rum.
SAM'L STEELE.

Fanny wants Salley to give Betty a little of the molases and Patsy Henderson some. I want you to send me one side of harness if you can get it without paying money. I don't care about it being very heavy, and send me a Tee-cattle and pay for it out of the wool, if you can get it at Waynes Borough.
JOHN McCLURE."

"7 August 1817, GREEN BRIER, VA.
DEAR SIR:
I have been wanting to hear from you and know what sort of a crop and what prospects of raising money for Bell this fall as I am in nead and what sort of a colt the sorl mare has and how everything is doing. I have rote to David McClure, that I hav a fine mare I wish him to take in Bells money. No more, only let me know how the meadows is.
SAM'L. STEELE.

Don't be so particular about not having anything to rite you certainly can always have something to rite."

The following letter is from Martha (Steele) Henderson, d. of James Steele and Sarah Wright:

"CALAWAY Co., MISSURIE, June 24th, 1830.
MY HIGHLY ESTEAMED FRIENDE:
I take my pen to write a few linese to you to informe you that we are all injoying good helth. Jane haste intirely recovered her helth. I have nothing particular to

wright to you, but that our friendship might live and not dye is my object in writing to you for you ever appeared to me as though you was a brother, it woode gratify me very much to here perticularly from you as we have never herde only Mr. Pilson wrote to Alexander that he thought that it would be very uncertain whether ever you woode come to this Country or not. I was afraid that when you was here, that you was not well enough pleased with the Country to come to live in it, though you had not a very good chance. I thinke that if you were here at this season of the year and had more time, you woode be better pleased, it would be a graite gratifycation to me if you ever moved anywhere that you would come to this part of the Country, altho I have never incoraged a friende that I have to come to this Country paste their owne inclination for I often wonder how it is that we are here ourselves, but I believe that whare ever people have to go to hav there bodyes deposeded there is always a meanes to take them there. I have often thought my Friend how much I lemented after my friends and how often I woode walke to the perere and looke as farre as I could see homewarde but oh I have got something to lement for now. I have often reflected on that, but my Friende these Providential thinges let it never be so harde with us we are not oblidged to give up to it. I was not so much surprised lately as I was when I herde of the death of two such blooming youths as your sister and cousin. I am deapely and senseably afected for all your loses and particular for youre aunte, who raised her from a childe and who felte to her as tho she was a daughter, but all that we have to comfort us in this life is that we would wish to hope, that our lose is gane to our neare and deare reletives. The Scripture tells us that we are not to morne, as those that have no hope. I muste conclude my letter. I send my love to you and cousin Jane, to youre uncle and aunte Fulton and your uncle Mitchel and aunte Mitchel and all the children and to Mr. Mathew Pilson and the reste of the family.

MARTHA HENDERSON.

P. S. Youre Uncle Mitchel watched the post office for some weeks after you got home for a letter. I woode like to have a letter from you."

The relation of Samuel Coursey, the writer of the following letter to John McClure, is not known. He must have been a kinsman, as Andrew McClure made his home with them:

"XENIA, OHIO, August the 11th, 1817.
HONORED SIR.

I now take this opportunity of writing to you to let you know that I have been well since I left that country; hoping when you receive these few lines you may be enjoying the same state of health, as for me to give you a full estimation of things I have seen during my absence from you I will not attempt, for I have not time, but when I come home then we will talk things over in full. I am highly pleased with this country and some parts of Kentucky, and much better pleased with some of the people. John, I tell you I think if you had come along with me out here we would both married before we would leave this country and in the best of familyes. There are one or two little girls here that seem to me that one of there names must be Coursey. John, I only wish you were here to go with me to the quiltings and visitings I have to attend to here with so many sweet little girls we would live fast I know. But stop, I forgot, how is them little Creek girls and Anny and Jane H. and Phebe and Jane P. and all the rest of the girls in that Country, are they all well. Say yes and tell them I am well to thank fortune * * * * I think you may tell the people that I am coming home soon, I am waiting on John Hutchison as I know he will be good company and he says I shant go till he is redy. Tell Sarah that I heard of some of her relations out here and will go and see them if I can possibly. Give my compliments to my father's family and Wm. Hutchison's family and John Diddle's family and all enquiring friends and so remain your friend and well wisher.

SAM'L LEWIS COURSEY."

William Berryhill was a cousin, son of Alexander Berryhill who m. Rachel Thompson April 6, 1786.

"BACHELORS NUNNERY,
GREEN COUNTY, OHIO, Aprile 2, 1820.

MR. JOHN MCCLURE.

Sir, after my best respects to you and yours, I drop you this line to let you know that I am still able to kick yet and have been kicking since I came home from Virginia. I have had my health tolerable well though somewhat unfortunate the last of February or the first of March. I had like to of got my leg broke by hawling shingle bolts in the waggon and throwing the bolts out of the waggon I had like to of threw myself out with one and to of broke my back. I thought the Bachelors Nunnery would soon come to a close for I thought I was to kill myself. But I have got able to kick again. I have put up a large barn, or at least it is 60 feet long by 25 wide and upward of 30 high, you may call it what sise you please, I have got everything redy to nail on the shingles. I want to put up a set of corn cribs as soon as I finish my barn, I have got the logs cut and this will be the fourth building that I have put up since I came home, or to the Bachelors Nunnery. I have and will have fifty or upwards acres of land opened and under fence this summer. I have had four hands working since last faul till last weak. I paid one of and sent him to work for his family, another went to bed last Sunday evening as usual and was a corp by three o'clock the next morning, his wife says she knew nothing was the matter with him till he was strugling his last. Tell Jane that I am coming to help her to eat that big cheese for I expect it will be a fat one and I am truly fond of good things and sweet things and true things. But sour things and faulse things I bid them good night. I suppose I might with propriety if report is true say Betsy child what are you doing. But I say keep a kicken for the blackest day has not come yet. I must leave this off and give you a history of our markett. Flour in Cincinnati is $3, whisky 40 cents, rye 37½, corn 25, wheat from 37½ to 50 cents, bacon 6¼ to 8 cents. I was offered several hundred of bacon for 6¼ cents. But I had no wife nor child to feed and in course I did not want to buy. I am yours with respect. WM. L. BERRYHILL.

Now, you will not forget to answer this line by the next post if you please and give me a full detail of all transaction that has occurd or taken place, since I left that part of the world. I expect to see you all by the first of August next, if I am spared. I have rote to Phebe and Mathew Pilson and have received no answer since. I suppose they think me not worthy of their attention. But you can tell them I am fat raged and saucy as usual and still true hearted and don't care for the purtiest——"

In addition to his farm of 699 acres in Augusta Co., John McClure at one time owned some land in Ohio.

"Know all men by these presents that I, John McClure of Augusta County, and State of Virginia, lease unto George Harmon of the aforesaid County, and State of Virgini, a tract of land lying on Deer Creek, Madison County and State of Ohio, for the term of one year, in consideration of which the said George Harmon obligates himself to make all the outside fencing, good staked and riders for the true performance of which we bind ourselves, our heirs, executors, or administrators in the penal sum of one hundred dollars. Given under our hand and seals this 10th day of September 1822.

JOHN McCLURE,
GEORGE HARMON.

Arch'd Stuart, Jr."

"State of Ohio, Madison County, 5th April, 1828.

Sir: I have not yet been informed, whether the contract between yourself and Mr. Dawson had been completed and having no instructions what to do, have on my own responsibility let both of the places out on the same term for the present year, that you rented them for: Tom Orpurd takes hoth. I have not received much of the rents yet. Samuel Houston died last fall was a year, his appraisers are unsettled. Ephraim Dawson died last Winter, his estate will not seen be settled. So that all I shall shortly receive will be what I get from Orpurd who has paid for the first year and says he will soon pay the second. I wrote to you (perhaps a year past) that a proposal had been made concerning the purchase of your land, the man who wanted to buy it has not yet purchased and still desires to hear from you concerning Terms &c. Your Friend.

JOHN ARBUCKLE."

"FAIRFIELD, July 31, 1832.
DEAR SIR.

Cousin Jane requested me to write to you when I returned home and let you know how she was. I left the White Sulphur on the 23rd; she was then as well as could be expected. The water had a tendency to sicken her a good deal when she first used it, but after a few days it had a better effect. There was a great crowd, 240 persons, and many new arrivals. My mother and her wish my father to start so as to be there by the 13th August; they were well situated and appeared to enjoy themselves tolerable well. You must not forget to write to her once a week as she requests it of you. She had not heard a word from any of you from the time she left home. I wrote to Matthew Pillson a few days before I set out, which news you have received. She said that she would be glad that Matthew would go out and spend a week with them. I would be glad if he could go. We are all well.

Yours respectfully,
A. ALEXANDER.

MR. JOHN McCLURE, Greenville, Va."

John Beaty was a friend and neighbor. The Beaty farm joining Old Providence Church, now owned by Samuel Finley McClure, was bought from the estate of John McClure.

"DUBLIN, WAYNE COUNTY, IA., Janr. 7th, 1832.
DEAR FRIEND.

At my father's request I write you a few lines. We all landed in Ia. the 19th Oct. The old man had to stop in

Richmond, about four miles in the State, partly on account of bad health and partly on account of bad roads and shackling teem. His bay filly failed before we got to Lewisburg and he swoped her off; we then got along tolerably till we came into the Ohio where the roads became bad. His old bay mare gave out & he had to buy one, otherwise we got along well. The rest of us went about 40 miles further and stopt in Henry County, the roads being so excessive bad that we thought imprudent to venture. Byers and myself then set out on horseback to visit the Warbash Country & went thro Marion, Boon, Mongomery & Clinton, & returned not satisfyed. We met with your brother. He removed from Tennessee and is keeping public house in Frankford, in Clinton County. We breakfasted with him. His family is well. I have settled about 18 miles west of Richmond in a small village and have purchased some property. The old man has purchased land within a mile & half of this place; has sold his waggon & two horses; the waggon for $100, the old bay mare that he got of Rowan for $50, the little pony for $40. He is now lying verry low and not much hopes of his recovery. He was verry much exhausted with fatigue when he stoped, but had recruited verry much untill the cold weather set in about the last of Nov. when he was taken suddenly ill with a relapse of his old complaint. Dr. Hindmad, of Richmond, is his physician, but does not entertain much hopes of his recovery. I was to see him a few days since; he was a little better, but has no hope, or but verry little, of ever being better. He requested me to write to you and let you known his situation. He wishes you to fetch your deed out with you, and also Alexander's and Thornton's and let William sign them all here, as he thinks it somewhat doubtfull whether Wm. can go in shortly if he should be called off * * * * Grain of all descriptions here is scarce and verry high, and of course, somewhat hard on emigrants. I believe I am thro. Give my respects to all my old neighbors. Tell Jas. Rowan that I will write to him shortly. My respects to your good lady and all the family.

While I remain your affectionate friend,

JOHN BEATY.

N. B. My Bob horse could not stand hard times, he quit eating a few days after I started and has hardly come to his appetite yet. During the trip he was compleaning every few days of the bots or cholic & was verry near loosing his eye sight. I have swoped him off.

J. BEATY."

In politics, John McClure was an old line Whig, strongly opposed to Secession, but when the crisis came gave his five sons to his State, two of whom paid the price in blood.

In appearance he was six feet, a man of great strength and endurance. In early life he made frequent trips to Scottsville and Richmond, marketing his produce, either driving his six horse wagon in person or accompanying on horseback his negro driver.

The following obituary written by his lifelong friend, Rev. Horatio Thompson, D. D., for nearly fifty years pastor of the Old Providence Church and one of the trustees of Washington College that elected Gen. Robert E. Lee to its presidency. "He filled a large place in the community giving moral tone wherever his shade was cast. A peacemaker, a benefactor—the poor man's friend and the idle man's dread. He was the Christian and gentleman of olden times—holding both sacred and honor bound. As he lived he died. We all say, a patriarch has fallen. He was a Presbyterian, and true to its code—a lover of all good without blushing to acknowledge it. A husband, father and grandsire, as devoted as these lofty names imply. He travelled to the tomb with manly bearing, where

> 'The trav'ler outworn with life's pilgrimage dreary
> Lays down his rude staff, like one that is weary,
> And sweetly reposes forever.'

H. T."

Of his wife JANE PILSON it was written:

"She connected herself with the Presbyterian Church of Tinkling Spring under the pastorate of the Rev. John McCue about the year 1816; removed her membership to the Presbyterian Church of Bethel, being a consistent member for near 66 years. She died in perfect peace in hope of a glorious immortality."

There were eight children born of this union.

(1). MARY MITCHELL. The following obituary appearing in the local papers at Gazelle, California, gives the outline of her life:

"Our last week's Journal chronicled the death of Mrs. Mary M. Harris. As a slight tribute to her memory, and an offering of esteem to her many friends and relatives, on this coast, as well as in the Eastern and Southern States, we will give a brief sketch of the life of this most exemplary woman.

Mrs. Harris was a native of Virginia; she was born June 12th, 1820, and passed peacefully away May 10th, 1892, aged nearly 72 years. She was the oldest of eight children, of whom but one, the youngest, survives her.

Miss Mary McClure was married when twenty years old, to Mr. Harris. In 1854 they moved to Illinois. Mrs. Harris was the mother of eight children, only five of whom survive her; four of this number are now residing in this country, near Gazelle, respectively, Mrs. E. B. Edson, John, Life and Susie Harris. Mrs. Harris accepted the bitterness of sorrow and loss with which her life was for many years overshadowed with patient fortitude.

Her's was one of those rare sweet natures, endowed with a Christ-like spirit, which shone out with a steady light, illumining her daily life of care with a mild radiance, almost pathetic in its gentle constancy. When her youngest child was but four years old, her husband suddenly dropped dead in the street, of heart disease, and after rallying from this first great shock of grief and affliction, she took up the burden of life with heroic grace. Her eldest child—she whom we know as Mrs. Life Edson—being at that time only sixteen years old.

Mrs. Harris raised all her family of eight, to be good and useful men and women. Three have since died, two being snatched away as suddenly as was their father. In addition to the care of her own family she raised two of her sister's children, a niece and nephew, their mother dying when they were very young.

The niece, Miss Stuart, was with her during the past year. She, with Mrs Harris' own children, were untiring in their devotion and care, doing everything in their power to make the last months of her life as comfortable as possible. She was always very happy with her children, who have all done credit to her careful training, and who awarded her to the full, that meed of honor and praise, justly due—if not in words—in that daily manifestation of love and respectful tenderness, so grateful to a faithful mother's heart.

Mrs. Harris had been a member, of the Presbyterian Church since early girlhood. Her's was a true Christian life, and she has no doubt found as many loving friends awaiting her in that bourne across the mystic river, as she has left here to mourn her loss.

With this assurance, let us not mourn, but rather rejoice with her in the sweet summons: 'Come up higher, thou good and faithful servant, thy lamp of life hath been kept burning, and thy way is not dark for thee. Enter now into the joy of thy rest.'

May 23rd, 1892. J. P. C."

She married Jan, 16, 1848, Thomas Harris, of Augusta County, Rev. A. B. McCorkle, pastor of Bethel Church, officiating.

Her eight children are:

a. Jane Elizabeth; m. E. B. Edson.

b. James, died in Iowa; m. Julia Page; left two children. (a) Minnie, who d. s., and (b) Jennie, who m. a Bancroft.

c. Sarah Margaret, b. 1844, m. 1772 Samual A. Lightner, of Augusta Co. She d. Oct. 17, 1873, and is buried at Bethel. The following is from The Telegraph, published at Dixon, Ia., where she lived before moving to Guthrie, Iowa: 'Died near Staunton, Va., Mrs. Sarah Lightner, wife of Samuel Lightner and daughter of Mrs. Mary M. Harris, of Guthrie, Ia., aged 29 years.

This sad intelligence brings sadness to a very large circle. So well known in this vicinity, her bright and lovely character rendered her a favorite among her friends. She was married last December at Guthrie, Iowa, and thence removed with her husband to Virginia, where she had a pleasant, happy home. For a number of years she was a member of the Presbyterian church at this place. She was an earnest, warm-hearted Christian, and she died as she had lived. Many hearts go out in sympathy to those dear friends who, in her loss have been so suddenly and deeply bereaved."

d. John McClure, living single on his ranch at Gazelle, California.

e. William Mason, m. Celia Sampson; died, leaving one son, Frank Sampson Harris.

f. Mary Susan, living single near Gazelle, Cal.

g. Eliphelet, farmer, living single near Gazelle, Cal.

h. Orpha Pilson, d. s.

(2). GEORGE WASHINGTON McCLURE, born Jan. 1, 1822. A conservative farmer and highly respected by his neighbors. He was less aggressive and successful as a business man than his father, but resembed him in his integrity and uprightness in the various relations of life. In personal appearance he was more like his father than any of the sons, standing six feet with broad shoulders, a man of great physical strength.

The following brief obituary notice appeared in the Central Presbyterian at the time of his death: "Died, Dec. 11, 1890, at his home near Spottswood, Va., after an illness of ten months with Bright's disease, Mr. George W. McClure, in the 68th year of his age. He had been a member of New Providence Church for many years and was one of our best citizens, and was beloved by all who knew him. 'An honest man, the noblest work of God.' "

Like his four brother, he was a soldier of the Civil War.

Having a wife and three small children, he did not enter the service at the beginning, but provided a substitute. He later enlisted, a private in Company H, 52nd Va. Regiment, and served until taken prisonor at Petersburg, Va., March 25th, 1865: In this battle he fought beside his brother James. Ordered to charge a battery, they reached the guns, one to die, the other taken prisoner. He was given permission to carry his brother from the field.

He married, first, February 28, 1850, Margaret Finley Humphreys, (October 5, 1829—March 27, 1870), daughter of Aaron Finley Humphreys, an Elder in Bethel Church. Five children:

a. Alexander Stuart, b. December 15, 1850, now of Staunton, Va. He m. October 26, 1882, Emma E. Moore, of Rockbridge County. Three children, viz.,—(a) Ida Margaret, b. August 9, 1885; (b) Kathleen Finley, b October 9, 1891; (c) Claudie Bell, b. March 30, 1893.

b. Jane Ann, b. August 25, 1853, lives near Spottswood, Va. She married (the second wife) October 25, 1881, Samuel A. Lightner, (1841-1904), son of Jacob Lightner and Mary Pilson. Four children, (a) Charles Thompson, b. January 20, 1885, and married, February 23, 1910, Bessie Wilson Ruff; (b) Frank Bell, b. January 5, 1887; (c) John Pilson, b. September 20, 1882, d. November 10, 1884; (d) Finley Alexander, b. January 20, 1893, d. July 27, 1893.

c. Mary Lou, b. April 25, 1857, and married November 18, 1880, Wm. H. Wade. They live near Brownsburg, Rockbridge County. Their children are (a) Finley Moore, b. October 24, 1881; (b) Geo. Wm., September 11, 1883, and m. June 10, 1908, Mary Maude Templeton; (c) John Edwin b. July 15, 1885, and m. February 2, 1909, Grace Berry Templeton, (d) Samuel Bell, b. March 1, 1887, m. December 30, 1908, Mary Ann Potter; (e) Zachariah Walker, b. December 8, 1888; (f) James Alexander, b. October 4, 1890; (g) Jacob Nevius, b. July 13, 1892; (h) Margaret Jane, b. August 6, 1894; (i) David Whipple, b. May 3, 1896, d. March 28, 1897; (j) Charles Thompson, b. February 1, 1898, d. November 5, 1900; (k) Robert McClure, b. March 13, 1901; (l) Frank Lightner, b. December 8, 1902; (m) Freddie Pringle, b. February 2, 1906, d. July 17, 1906.

d. John Finley, b. September 1, 1861, and died February 4, 1906. He was a successful farmer, and one of the best business men of his community. He was from early youth a member and for several years before his death an efficient Deacon in New Providence Church. He married November 28, 1894, Anna Poage Willson, b. March 9, 1864, daughter of John Edgar Willson and Elvira Ann Brooks, and sister of Colonel James W. Willson, Superintendent of the New Mexico Military Academy, Roswell, N. M. He bought soon after his marriage the farm near Fairfield, Va., formerly owned by his wife's father, where he died of asthma in the 45th year of his age. Their children are, (a) George Edgar, b. October 26, 1896; (b) Elvira Brooks, b. October 9, 1898; (c) James Finley, b. July 16, 1901; (d) Finley Willson, b. October 17, 1903.

e. Sallie Bell, born June 3, 1864, married December 8, 1885, Frank M. Oates, (1863-1911), of Pope County, Arkansas, where she died June 3rd, 1887.

George W. McClure married, second, March 19, 1872, Susan Rebecca Foutz, born May 15, 1850, daughter of Henry T. Foutz and Mary Jane Craver, of Rockbridge County. Four children:

f. Clara Steele, b. January 2nd, 1873, and married (the second wife) April 4, 1899, John W. Martin, of Nelson Co., Va., who died February 20, 1910, aged 61 years. Issue (a) Elnora, b. January 11, 1904.

g. Minnie Merle, b. December 29, 1877, is living single, Spottswood, Va.

h. Clay Pilson, b. December 17, 1879. Has bought the farm his father owned, where he is engaged in farming and stock raising.

i. Lillie Luster, b. June 2, 1885.

(3). SARAH STEELE MCCLURE, b. Feb'y 23, 1824, d. April 9, 1873.

The following obituary was written by Rev. Horatio Thompson, D. D., pastor of Old Providence Church:

"Died, April 9th, in Augusta county, Va., Mrs. Sarah

S., consort of the late Andrew Stuart, in the 50th year of her age.

It is an instinct of nature to recall the past. As we draw the veil aside, blessed memories rise up before us. Now, joyous in the bygone—now sad, that they are gone forever. To name the subject of this obituary fills the heart with emotion and the eye with tears. She was all that we mean when we speak of woman in her loveliest sense; a Christian without pride, a wife without discontent, a mother without a frown, a neighbor without an enemy. To forgive was the attribute of her nature; to make every one happy was a law of her heart. As might be expected, she was loved and mourned by all.

Her disease was consumption which, though protracted, was tenacious of its claim. At the decease of her amiable husband, she, with two orphan children, were taken into her parents' house, where each member bore a part of her grief, and brothers and sisters drew around her with double affection. But there she must not stay. She is hurried from a kind father's house to the home of a heavenly Father. Now it is that afflictions join hand in hand. The father and daughter enter upon the same pilgrimage as if by agreement. Daily they exchange calls with feeble step till flesh and heart fail and she sleeps to wake no more.

Two lovely children mourn her loss and numerous friends and neighbors attended her to the tomb. She was a member of the Presbyterian church at Bethel, and there rests from every toil. Here's is eternal fruition—ours to mourn. 'Even so, father.' "

She m. Andrew Alexander Stuart, son of Archibald Stuart and Polly Alexander.

Two children,—

a. John Thompson, m. Sadie McGilvray, of Richmond, Va. One child, John T. Jr.; d. i.

b. Mary Steele, m. William T. Hutchinson, of Rockbridge Co. One child, Mary Stuart.

For the Stuart Family see Waddell's Annals of Augusta Co., p. 366.

MONTGOMERY McCLURE,
1825-1892.

(4). ANDREW WELLINGTON McCLURE, b. May 19, 1826, and died at his home (now the residence of Rudolph Bumgardner), 205 N. Augusta street, Staunton, Va., Feb'y 12, 1878. Buried at Bethel. Of the firm of Bumgardner & McClure, he was for a number of years a merchant at Greenville, and later at Staunton, Va. He was a soldier in the Civil War, 10th Va. cavalry, Capt. Ed. Fulcher's company, Gen. Beale's Brigade. Dick, his war horse, lived for many years after the war, showing a bullet wound in the neck.

In personal appearance he was 6 ft. 2 in. in height, a strikingly handsome man.

He married April 5, 1853, Mary Bumgardner, (Aug. 9, 1836—Oct. 20, 1884), oldest d. of Lewis Bumgardner and Hettie Anne Halstead.

Ten children—

  a. Phoebe Jane, a lifelong invalid, b. Oct. 31, 1855.

  b. Malinda Halstead, b. August 22, 1858, m. Warren Case, now deceased, and lives in Jacksonville, Ill. Two children: Warren and Mary.

  c. Mary Stuart, b. March 9, 1863, m. John H. McClure. Four children.

  d. Frank, d. i.

  e. Alice Clara, b. Apr. 17, 1865, m. Horace Bougere, of Louisana. Two children living: Ethel and Carl.

  f. John Andrew, d. i.

  g. Hettie Anne, b. March, 1870, m. James Capps, of Jacksonville, Ill. One child, James Capps, Jr.

  h. Sarah Steele, b. Sept., 1872, m. William Wilcox and lives in Birmingham, Ala. Four children: William, Frederick John, Malinda and James Gallaher.

  i. Andrew Wellington, Jr., d. i.

  j. Katie Wellington, b. Feb. 7, 1877; m. June 10, 1903, Richard C. Reynolds, of Jacksonville, Ill.

(5). JAMES ALEXANDER McCLURE, born Oct. 21, 1828; owned a farm deeded him by his father near Spottswood, Va., where he was living at the beginning of the Civil War. He was mortally wounded in a charge on Fort McGilvray.

The story of his life has been written as follows: "James A. McClure, of Augusta Co., Va., fell mortally wounded in the battle of March 25, 1865, near Petersburg, Va.

Brought up by pious parents in the ways of wisdom and that virtue which ennobles our nature, his whole life was a verification of the promise made to those who train up their children in the way they should go; and like the young man our Saviour loved, it might almost be said of him, that he kept all of the commandments from his youth up. In the formation of a character so correct as to be almost faultness, the governing principle was not fear, but love—the love of God, of his Saviour and of virtue. Consequently when duty called he hesitated not to go, though it were to the mouth of the cannon where he fell.

In every station and relation in life he acted well his part where all the honor lies; but beyond all this there were distinctive characteristics that marked the man and caused him to be respected and beloved by many of his fellows. His disposition was frank, cheerful and happy; his benevolence disinterested, his generosity whole-souled and free, while meanness was an utter stranger to his nature. To him there was a luxury in doing good; and to crown all his faith in Christ was a single childlike simplicity, beautiful to him who witnessed it, and exciting the feeling 'of such is the Kingdom of Heaven.'

With such a character, it is not necessary to say that he was all a neighbor, a friend, a father, a husband could be. It was in these relations that his modest unobtrusive virtues shown most brightly and here the blow falls with the most crushing weight."

The following from the Minutes of Mt. Carmel Session.

"The Session of Mt. Carmel Church, on receiving the mournful intelligence that James A. McClure, a member of this body, had fallen on the battlefield mortally wounded, and soon after expired, in order to express their high appreciation of his worth, and pay an humble tribute of respect to his memory, adopted the following minutes: Whereas, It has pleased our Heavenly Father to remove by death one of our members, James A. McClure, therefore resolved,

1. That we accept of the affliction as coming from God, and bow submissively to His righteous and sovereign will.

2. That in his early death this congregation has lost an efficient office-bearer—a useful and beloved member.

3. That we tender our warmest sympathies to the family and friends of the deceased, and commend them to Him whose grace can soothe and heal the broken heart.

4. That a copy of the above be forwarded to the family and published in the Central Presbyterian.

JAMES HENRY, Clerk of Session."

He was ordained a ruling elder in the Mt. Carmel Church at an early age.

He entered the Confederate service early in the war as a private in Co. H, 52nd Va., Infantry, in which service he gave his life, March 26, 1865.

The following letter is from his brother-in-law to the father, John McClure, Greenville, Va.

"SAM STEELE'S, April 2nd, 1865.

ESTEEMED FRIEND.

I write to inform you that we have just heard from George and James through James T. Black, who wrote to his wife who is in Richmond on a visit to her brothers and she writes to her friends, that they are both prisoners in the hands of the enemy, and that James is badly wounded—having his thigh broken. We have not heard what part of the thigh or any particulars. You will please inform Cousins Rebecca and Margaret.

We are tolerably well and will go up this week if not disappointed in getting a horse.

Very respectfully,

A. A. STUART.

P. S. They were taken prisoners in the fight of the 25 of March."

He married, April, 1853, Rebecca Hemphreys, (May 19, 1832-September 3, 1896) d. of Samuel Humphreys, and sister of Caroline, wife of Robert Tate Wallace, Rev. James Humphreys, Dr. William Humphreys, Captain John Humphreys, and Mrs. Jane Donald. Two children:

a. John Howard McClure, b. June 30, 1854, now living near Brookewood, Augusta Co. where he is extensively engaged in farming and stock raising.

He married Mary Stuart, daughter of Andrew Wellington McClure, Sr., and Mary Bumgardner. Four children.

(a). Hugh, b. March 12, 1889, student Virginia Military Institute. Bank Clerk, Staunton, Va.

(b). Edward Donald, b. May 3, 1891, student Washington and Lee University. Merchant, Spottswood, Va.

(c). Reba Bell, b. May 17, 1893.

(d). Mary Alice, born Jan. 14, 1901.

b. Samuel Finley McClure, b. Feb. 10, 1858.

He lives at Spottswood, Va., of the firm of Spencer & McClure, where he owns a large farm, and is one of the most honorable and successful business men of his section of the county.

He was while quite young made a deacon in Mt. Carmel congregation, and at present is one of their most efficient Ruling Elders.

He married, first, Anna McChesney, who died leaving no children.

He married, second, April 29, 1908, Mayme Smith, d. of Edward Lewis Smith and Clara Weir, both of Rockbridge County. Two children.

(a.) Samuel Finley, Jr., b. June 24, 1909.

(b.) Jean Weir, b. January 3, 1914.

Edward Lewis Smith was born in Rappahannock County, the son of Oliver Perry Smith and Margaret Massie.

Clara Weir was the daughter of John Weir and Margaretta Brooks, the latter a daughter of Eleanor Mayberry, of Philadelphia.

(6). JOHN PILSON McCLURE was born Apr. 9, 1831, and died at his home in Augusta County Jan. 3, 1865, from the effects of an accidental wound. Like his father and brothers he was a farmer, and was settled in his home at the beginning of the war. The following obituary notice, published at the time of his death and written by Rev. Horatio Thompson, D. D., gives in brief the story of his life:

"At a time when the harvest of death is so abundant and the young men of the South are falling by thousands, it may be thought unnecessary to chronicle their names, but every true soldier is a *star*, and his position in the social and national firmament is important, though 'one star may differ from another star in glory.' As such, posterity claims the right of fellowship with the gallant dead, when the drama ends, and naught but the soldier's name remains upon the scroll of his country. Hence we record for the future reader the name of *John P. McClure*.

Nothing brilliant accompanied his chilhood and youth except a strict tenacity for truth and honesty. It was reserved for the last years of his life to develop the true patriot in his full proportions. He was Southern to the death, and truer steel was never hilted, nor more resolutely wielded, for it was tempered in the fire of patriotism.

When the war broke out he was pronounced unfit for service from an oganic affection of the heart, but in the second year of the war he believed his services were demanded and volunteered in the 14th Va. Cavalry operating in Western Virginia. All hailed him as a valuable accession, especially the members of his own company, the 2nd Rockbridge Dragoons. In this command his moral worth was most salutary and was only equalled by his bravery. Finding the service too severe for his constitution, he procured a substitute and returned to his welcome home. (Here his influence was also salutary, as who will not discover upon entering that little circle where a lovely wife and four children meet the visitor with that subdued demeanor and gentle smile, which tells the story of a deceased husband and father).

But Congress soon ordered to the field all who had placed substitutes in the ranks and again home is fled and the soldier's armor girded on. He entered the 23rd Va. Cavalry, where he continued till the time of his death, which was occasioned by an accidental wound terminating upon the lungs.

Owing to his bravery he was employed in scouting and detached service by his Generals.

At the battle of Piedmont his horse was shot, and near Staunton, as General Hunter in his memorable raid passed through the Valley, he was foremost in a charge and captured several prisoners. He afterwards entered the enemy's camp making discoveries of importance. He was one of a small party of scouts who gladdened the citizens of Timber Ridge and Lexington on the morning of the 14th of June, filling all hearts with joy as the Confederate war-horse and gallant soldiers wound their way along the gloomy waste, and charging into Lexington captured several prisoners while a Yankee regiment was yet in town. As the party rode up to our door we supposed they were 'Jessie Scouts,' and asked if he were a prisoner. He said he was not, and that they were Confederates. Such intelligence after such suspense filled us with emotion.

But there was something in this good citizen and soldier which exalts him above the praise of men. He was a Christian. In early youth he gave himself to Christ and united with the Presbyterian Church at Bethel. He was cheerful even in war and hopeful under discouragements. He was affectionate and devoted to parents, who yet live to cherish his memory and recount his filial offices. As a husband, he was all that is embraced in that hallowed word. As a father, he was tender and loving, yet manly and dignified.

Death came upon him suddenly, though ready for his coming. He was first to discover that death was approaching, and being asked if he was afraid to die, said 'Oh no, I have long lince given myself to God; 'O God, thou art my Rock, in Thee will I put my trust.' Then bidding all pres-

ent farewell, he said 'Lord Jesus, receive my spirit.' He was universally beloved. A large connection mourn his loss, but especially those who bear the widow's grief and weep the orphan's tears. But this promise is theirs, 'I will be a husband to the widow and a father to the orphans.'
H. T."

The funeral was preached by Rev. Wm. Pinkerton, pastor of the Mt. Carmel Church from Ps. 37:18, "The Lord knoweth the days of the upright and their inheritance shall be forever." His grave is marked in the family plot at Bethel Church.

It will be of interest to his posterity to know that with the exception of the organic trouble referred to above, he was a magnificent specimen of manhood, standing six feet two inches, and his unusual feats of strength are still remembered by those who knew him in his early manhood. He was a most delightful companion, with a sense of humor and a rare gift of wit that was more than once a means of grace. Soldiers of his command still relate how in the most terrible hour of battle, in the most dangerous situations, he would make some droll, unusual remark that would provoke a peal of laughter and send men into the charge of battle with a smile.

He married February 14, 1856, Mary Tate Wallace, daughter of Robert Tate Wallace and Caroline Humphreys, two families prominent in Augusta Co., from its earliest settlement. She survives with their four children, viz.

a. Robert Wallace McClure, b. May, 14, 1857. He was for a number of years a public officer in Augusta Co.; one of the first deacons ordained 1873, and treasurer of Bethel Church, and at present one of its leading Elders. He lives near Greenville, Augusta Co., where he is a prominent farmer. He married November 22, 1893, Ada Brubeck, the only child of Jacob Brubeck and Essie Ott. Four children:

(*a*) Virginia Wallace, b. October 2, 1894.
(*b*) Robert Vance, b. November 22, 1895.

(c) Mary Mildred, b. September 23, 1897.

(d) Essie Ott, b. April 15, 1899.

b. John Marshall McClure, b. April 14, 1859, and was named by his grandfather McClure for Chief Justice John Marshall whom he greatly admired. True to the best family tradition he also is a prosperous farmer and lives about three miles north of Old Providence Church.

As a ruling Elder, he has for a number of years been prominent and useful in the large Bethel congregation.

He married November 6, 1890, Linda Sproul, daughter of Archibald Alexander Sproul and Engenia Bumgarder, a great granddaughter of Rev. Archibald Scott, the Revolutionary War pastor of Bethel congregation. She died after a few years. He married second, December 3, 1912, Mary Scott Storey, d. of Rev. G. T. Storey, of Houston, Texas.

c. Charles Dorsey McClure, b. April 13, 1861. He now owns and lives on the McClung farm adjoining that of Marshall McClure, and operates the old McClung Mill.

He married Ella Mish, a daughter of James Smith and Cornelia Wallace and widow of the late George Mish. Six children:

- (a). Mary George, b. February 3, 1889, and died June 1, 1890.
- (b). Carrie Louis, b. August 25, 1890.
- (c). Edna, b. February 21, 1892.
- (d). Lucy Moore, b. February 9, 1894.
- (e). Charles Dorsey, Jr., b. September 15, 1902.
- (f). Russell Carlyle, b. November 6, 1904.

d. Carrie Pilson McClure, b. August 3, 1864. She m. June 29, 1892, James Scott Callison, an alumus of Washington and Lee, and the University of Virginia. He is a useful and influencial citizen and a valuable and efficient member of Bethel church. Three children:

- (a). Mary Wallace, b. August 13, 1893, alumna Mary Baldwin Seminary and Converse College, S. C.
- (b). James Waller, b. March 1, 1895.
- (c). Marshall McClure, b. November 9, 1896.

(7). ELIZABETH, b. June 5, 1833, d. July 23, 1833. Buried at Bethel.

(8). MATTHEW THOMPSON McCLURE, named for a great grandfather, born July 23, 1834. He lives at the old McClure home where he and all his brothers and sisters were born, and where his father and grandfather died, one mile northeast of Old Providence Church. He is the last of his generation. Like his father and grandfather and great grandfather he has lived to be eighty years old.

He has for many years been looked upon as one of the most intelligent and successful farmers of his section. Without a scientific education, he has through years of experience mastered many of the facts of agricultural chemistry to the advantage of his community. Even the very briefest sketch of him would be imperfect that did not mention his keen sense of humor that has made him throughout his life a delightful companion and welcome guest; a hopeful disposition with an indomitable will and a strong character that has ever enabled him to rise triumphantly above the misfortunes of life.

He has served in a number of positions of trust and honor in his community, such as over seer of roads, school trustee, chairman of his district party (Democratic) organization. Urged by friends to become a candidate for the Legislature he resolutely declined. Along with two other gentlemen, J. M. Harris, and Baxter Rowan, he founded, and fostered for several years, a Classical School at Old Providence that was of untold benefit to numbers of young men and women, who would never otherwise have been prepared for College or secured a High School education. He has through a long life labored for every interest of his community, religious, educational, political and commercial.

Like his four older brothers he was a soldier in the Civil War.

The following papers bear on this period:

"Camp 52nd Virginia Infantry, Dec. 16, 1863.
To all whom it may concern.

The bearer hereof, M. T. McClure, Commissary Sergeant of the 52nd Va. Regiment, aged 29 years, 5 ft. 9 inches high, fair complexion, blue eyes, brown hair, and by profession a farmer; born in Augusta Co., State of Virginia, and enlisted at Staunton, in the County of Augusta, on the 31st day of July, 1861, to serve for the period of 12 months, is hereby permitted to go to his home in the County of Augusta, State of Virginia, he having received a furlough from the ——day of —— to the—— day of—— at which period he will rejoin his company or regiment at Somerville Ford or wherever it may be, or be considered a deserter.

Subsistence has been furnished the said Serg. M. T. McClure to the 16th day of December, 1863, and pay to the 31st day of October, 1863, both inclusive.

Given under our hand at Camp 52nd Va. Reg't, this 16th day of December, 1863.

   G. M. COCHRAN, 52nd Va. Reg't.

The reason Sergeant M. T. McClure desires a furlough, is that he has just heard of the extreme illness of his mother in Augusta County, Va., whom he desires to visit. The applicant's character as a soldier has been unexceptionable.

   G. M. COCHRAN, Capt.

This application was endorsed as follows:

"52nd Virginia Regiment, Dec. 16, 1863.

The applicant has never received transportation to his home and back under the law of Congress when on furlough of Indulgence.

Resp'y forwarded and approved.

   G. M. COCHRAN, Capt. 52nd Va. Reg't."

"Hd. Qrs 52nd Va. Reg., Dec. 16, 1863.

Respectfully forwarded and approved.

Sergt. McClure has always been a most faithful and exemplary officer.

This extraordinary application is submitted under the painful circumstances of the extreme illness of his aged mother, of which he has just received reliable intelligence.

JAMES H. SKINNER,
Col. Com'ng 52nd Va. Reg't."

"Hd. Qrs Pegram's Brig., Dec. 16th, 1863.

Approved and respty forwarded.

JOHN PEGRAM, Brig. Gen'l."

"Head Qrs Early's Division, Dec. 16, 1863.

Respectfully forwarded and approved for 15 days to be counted as part of the regimental quota under G. O. 84.

HARRY T. HAYS, Brig. Gen. Comd'g."

"Hd. Qrs 2nd Army Corps, Dec. 17, 1863.

Respectfully forwarded and approved under Gen. Hays' endorsement.

R. S. EWELL, Lt. Gen'l."

"Hd. Qrs. A. N. Va., 17th Dec., 1863.

Respectfully returned as recommended by Gen'l Hays. By order of Gen'l Lee.

W. H. TAYLOR, A. A. G."

Entering the Confederate service as Com. Sergeant in Co. I., 52nd Va. Infantry, made up of men of Augusta Co., Sam'l A. Lambert, Captain; John B. Baldwin, Col., they were first attached to Gen. Edward Johnson's Brigade, afterwards com'd by Gen. Early, who was succeeded by Gen. Pegram. Stonewall Jackson's Corps.

He was on Nov. 23, 1861, commissioned Second Lieutenant:

"The Commonwealth of Virginia.

To M. Thompson McClure—Greeting:

Know you That from special trust and confidence reposed in your fidelity, courage and good conduct, our

Governor in pursuance of the authority vested in him by the Constitution and Laws of this Commonwealth, doth commission you a Second Lieutenant of Light Infantry in the 93rd Regiment of the 13th Brigade and 5th Division of the Virginia Militia to rank as such from the 23rd day of November, 1861.

In testimony whereof I have hereunto signed my name as Governor and caused the seal of the Commonwealth to be affixed this 5th day of February, 1862.

JOHN LETCHER."

On the reorganization of the army he was retained as Commissary Sergeant Co. I, in which capacity he served until the close of the war.

Early in 1865 he was made First Lieut., with rank of captain, the office being reserved for either Capt. James Bumgardner or Capt. John Humphreys, prisoners, in case either should be exchanged. The surrender came before he received his commission. He has preserved his parole, which is as follows:

"APPOMATTOX COURTHOUSE, VA., April 10th, 1865.

"The Bearer, M. T. McClure, Com's'y Serg't of 52nd Va. Regt. of Early's Brigade, a Paroled Prisoner of the Army of Northern Virginia, has permission to go to his home and there remain undisturbed. He is entitled to take with him one private mule.

S. W. PAXTON, Capt.,
Com'd'g 52nd Reg't Va. Infantry."

Pocketing his parole he mounted his little mule with an old blanket for a saddle (his saddle was stolen the night before), and set out for home. Leaving Appomattox in company with Capt. S. W. Paxton and Dr. John Gibson, they spent the first night at Lynchburg, proceeding up the canal they spent the second night at Riverside with the Shields, the third night near Fairfield with Capt. Paxton, reaching home on the afternoon of April 14.

He rented his father's farm, which he later partly inherited and partly bought, and where he now (1914) lives, a worthy and honored citizen.

He married July 27, 1865, Sarah Catherine Bumgardner, b. in Carroll Co. Mo., March 18, 1842; daughter of Lewis Bumgardner and Hettie Ann Halstead, Rev. Francis McFarland, D. D., officiating.

It was a home building indeed, the family had been living in cramped quarters since the loss by fire of the substantial and commodious brick structure built in 1844, and no sooner settled than he erected the present brick building, inferior in size and appointments to the former one, but the scene of a happy home, and with the exception of the oldest, the birthplace of their nine children, as follows:

a. Lewis Bumgardner McClure, b. in Greenville, Va., Feb. 12, 1866.

After a good business education, secured at a classical school at Tinkling Spring and at Dunsmore's Business College, he located in 1887 in Russellville, Arkansas. He has been for a number of years Cashier of the People's Exchange Bank of that place, enjoying the confidence and the esteem of his community.

He was ordained an Elder in the Southern Presbyterian Church of Russellvill, April 6, 1892. Was a commissioner from Washburn Presbytery to the General Assembly of the Presbyterian Church in the United States, Lexington, Va., May, 1903.

Married May 8, 1890, Allie Bayliss, d. of Benjamin Franklin Bayliss and Sarah Evants, both of Pope County, Arkansas; the former a Confederate soldier, and at the time of his death in 1886, was clerk of the Circuit Court of his County. His father, Andrew Jackson Bayliss, moved to Arkansas from Tennessee in 1836. He was one of the pioneer school masters in that part of the State. He later practiced law, and for many years served as a clerk and as Judge of the County and Probate Court. He personally preserved the records of his county from destruction during the Civil War by removing them to the Ozark Mountains where he concealed them in a cave, bringing them back after the close of the war. Five children:

(a). Mary, b. July 19, 1891, and d. July 21, 1891.

(b). Lewis Bayliss, b. Nov. 16, 1892, and d. Jan. 14, 1894.

(c). Benjamin Thompson, b. Jan. 28, 1895, student at Washington and Lee University.

(d). Thomas Bumgardner, b. Aug. 7, 1896.

(e). John Bayliss, b. Mar. 12, 1900, and d. July 14, 1900.

b. Jane Thompson McClure, born April 2, 1869; m. Dec. 27, 1902, Edwin Bumgardner. They live at Walnut Grove, the old McClure home, the comfort of her parents declining years. She exemplifies in her energy, unselfishness and ability, the very best traditions of her people; a rare and noble woman. By her energy and unselfishness made a college education possible for four of her brothers.

c. Anne Halstead McClure, b. Sept. 28, 1870, and m. Oct. 4, 1893, Samuel Walter McCown, b. April 4, 1854, son of John Kinnear McCown and Mary Wilson; a ruling Elder in New Mammouth Church, as was his father before him. A conservative farmer and one of Rockbridge's most highly respected citizens. Their children, unusually attractive, are:

Mary Wilson and Katherine Bumgardner, twins, b. Oct. 1, 1894.

Agnes Stuart, b. Sept. 30, 1895.

Margaret Jean, b. Jan. 25, 1897.

Samuel Walter, Jr., b. June 22, 1898.

Sarah Jaquiline, b. May 9, 1900.

Halstead McClure, b. Aug. 29, 1901.

Katherine, a child of brightest promise, and of unusual gifts of grace and nature, died of typhoid fever August 29, 1909.

d. James Alexander McClure, b. Dec. 12, 1872. Student W. and L. University and graduated A. B. in June, '99. In The Calyx (W. and L. Annual) 1899, p. 19, we read, "James Alexander McClure, 'Yon Cassius hath a lean and hungry look.' He is genuine Scotch-Irish. In '96 was Final

Orator of Graham-Lee Literary Society, and in '99 Debater at the Intermediate Celebration of same society; member of Ring-tum Phi Staff '99, and Vice-President Y. M. C. A. '99; Valedictorian Class '99. Will enter the Christian ministry." The Southern Collegian, June, 1899, page 233:" The valedictory address was delivered by Mr. James A. McClure, of Virginia. Mr. McClure's speech was excellently written and delivered, and the latter portion, taking leave of Lexington and its fair ladies, was full of humor and kept his audience convulsed with laughter." P. 274, "Mr. McClure captured his hearers at the start, and, by the time he had reached the quarter-stretch he had them 'going all his way.'"

Entered Union Theological Seminary, Richmond, Va., Sept., 1899, and graduated B. D. 1902. In his senior year was Editor-in-Chief of The Union Seminary Magazine.

Entered Princeton Seminary as a gradute student 1902, and graduated B. D. in 1903. Pastor Second Presbyterian Church, Petersburg, Va. Author of "The McClure Family."

Mar. Dec. 31, 1903, at the home of the bride, Rev. R. A. Lapsley officiating, assisted by Rev. C. F. Myers, Josie Charlton Gilkeson, d. of John William Gilkeson and Margaret Letitia Tate. Five children:

(a). Margaret Randolph, b. Front Royal Va., Feb. 24, 1905; baptized by Rev. Chas. F. Myers, May 28, 1905.

(b). John Gilkeson, b. Buena Vista, Va., Nov. 5, 1906; baptized by Rev. A. H. Hamilton, Jan. 4, 1908.

(c). Jane, b. Henderson, N. C., July 29, 1907; baptized by Rev. H. B. Searight, March 19, 1909.

(d). Katherine Bumgardner, b. Henderson, N. C., April 13, 1910; baptized by Rev. Alexander Sprunt, D. D., May 24, 1911.

(e). James Alexander, Jr., b. Petersburg, Va., Aug. 4, 1912; baptized by Rev. C. R. Stribling, April 6, 1913.

e. Andrew Wellington McClure, b. Dec. 25, 1874. After a full commercial education he filled positions in Richmond, Roanoke and Staunton, Va., and in Bristol, Tenn. In this last named city he was ordained an Elder in the Central Presbyterian Church. Now located in Macon, Ga.

He m. Oct. 5, 1904, Julia Elizabeth Deyerle, d. of Henry Deyerle and Sarah Price, of Rocky Mount, Va., an alumna of Hollins Institute. One child:

(a). Sarah Katherine, b. Jan. 25, 1907.

f. William Warren McClure, b, Nov. 26, 1877, an alumnus of Washington and Lee University and Dunsmore's Business College. He emigrated to the West in 1897, filled positions in N. D. and Washington, finally settling in Lewiston, Idaho, where he now lives.

He m. Nov. 7, 1906, Anna Edith Day, born near Albany, N. Y., July 26, 1879. Her father, Wm. Day, was born in Wedmore, England, Dec. 25, 1851, and came to New York State, 1870. Her mother, Elizabeth Cox, was born of English parents in Glamorganshire, Wales, April 11, 1859, and emigrated to New York State, 1871. Wm. Day and Elizabeth Cox were married near Albany, N. Y., in 1876, moved to California in 1880, and to Washington State 1883.

Children,—

(a). Willam Thompson McClure, b. Aug. 18, 1907.

(b). Lewis Day McClure, b. Sept. 25, 1910.

(c). Georganna McClure, b. Aug. 6, 1912.

g. Mary, b. Aug. 11, 1879, d. Sept. 6, 1879.

h. John, b. Dec. 16, 1880. A. B., Washington and Lee University 1904. Graduate student Chicago University. 1908 and 1914. Professor of Chemistry in New Mexico Military Institute, Roswell, New Mexico.

He married June 18, 1913, Caroline Matilda, dau. of Mr. and Mrs. G. Frederick Fordon, of Geneva, N. Y.

i. Matthew Thompson McClure, Jr., b. April 27, 1883. Student Washington and Lee University, 1900-'05; student University of Virginia, 1906-'07 and 1909-1910; student Columbia University, 1910-'12.

Degrees: B. A., Washington and Lee, 1904; M. A., University of Virginia, 1907; Ph. D., Columbia University, 1912.

THOMAS McCLURE,
1795-1870.

Principal Louisa High School, Louisa, Virginia, 1907-'09; instructor in philosophy University of Virginia, 1909-'10; assistant in philosophy Columbia University, 1910-'11; University Fellow at Columbia, 1911-'12. Is now (1914) teaching philosophy at Columbia University.

Was debater and medalist, Graham-Lee Literary Society W. and L. U., 1901.

Member of the Phi Beta Kappa, University of Virginia, and Theta Delta Chi fraternity, author of "A Study of the Realistic Movement in Contemporary Philosophy," and joint author of "Guide to the Study of the History of Philosophy."

5. THOMAS McCLURE, youngest child of Andrew and Mary Mitchel McClure, was born near Waynesboro, Augusta County, Va., August 11, 1795, named for his grandfather, Thomas Mitchel, raised by his aunt, Elizabeth Mitchel Fulton, his mother having died when he was three weeks old. The following letters, now in the hands of the writer, together with a newspaper notice of his death, give us the outline of his life:

"CLINTON COUNTY, IND., June 18, 1830.
Dear Brother and Sister.

I gladly embrace the preasant oportunity of informing you that we enjoy tolerable degree of health, for which reasons we feel thankful to God for His mercies towards us and sincerely hope these lines will find your family enjoying the like blessing with my good old uncle and Ant Betsy, whom I oftimes have desired to see and converse with face to face and have prayed that our Heavenly Father would comfort them in their declining years with the riches of His grace, and as their bodies ripen for the grave, their souls may be ripening for heaven and glory beyond this vale of tears. My oald friends contend for the promise they that prove faithful till death shall have a crown of life at God's right hand, for our lives is frail and uncertain. See the instance of our dear sister, who was a few months agow in good health is now in eternity. I suppose you are anxious to know what I am doing and how I am doing. I am raising a crop on the twelve mile perarah near the Wabash in the above named county, one mile from where the county seate is located, where a sale of Lotts will take place in July when I am desirous to seat myself for my business, for there is a great opening at preasant and the sooner I embrace the opertunity the better. It has the prospect of being a very rich country. It is settling very fast and with men of wealth. My staying so long from my family is very painful to me, but my misfortunes has forced me to try to make something for my family. I have above 30 acres in corn and oats and looks equal to any ever I saw. I think I shall have grain in abundance and grain is very high yet on account of people mooving so fast to this country. I design gowing back to Russell and Scott in July or first of August to try to settle up my affairs. I would be very glad to see yon onst more, but it is impossible for me to get to that country this season and move, for I am determined to move to a new country where I can have some chans to get a good home. Give our Love and respects to all enquiring friends. This from your loving Brother till death.

THOS. McCLURE."

"CLINTON COUNTY, IND., July 18, 1830.

DEAR BROTHER:

I again take up my pen to adress a fieu lines to you, least you should not get the first, and say through mercy I am enjoying health, but my companion has been very unwell for near 3 weeks past and recovers slowly. My crop looks well. I think I shall, if the season is good, have upwards of 12 hundred bushels of corn agreeable to the calculations amongst the people. I have betwixt 4 and 5 acres of the best oats ever I saw. Their is all the encouragement for people to work that is necessary and every thing bears a very fair price; a fine wheat country. We had a sail of Town lots on Monday last; Lots soald High and ready. I purchased one of the first chois lots in town. We have three payments, equal one in hand two annually. All I lack of settling myself is a little capital; people gave me all the encouragement I could ask. I intend starting in to Russell and Scott about the first of August to settle my business and moove my children. I feel distressed in mind about the situation of my family, being separated, but I am trying to do the best I can for them. I desire greatly to see you and rote to you to meet me at James' in Lebenon about the first of September. Remember my love to all enquiring friends. This is from your loving Brother until death. THOS. McCLURE."

"CLINTON COUNTY, INDIANA, Nov. 19, 1838.
DEAR AUNT.

I received your letter by cousen T. C. Mitchel, which gave me much satisfaction to hear you were still in the land of the living enjoying tolerable health and alsow found us enjoying the like blessing and feel thankful to God for His goodness. I still have it in contemplation to pay you a visit as soon as circumstances will admit of it. As to my having forgot you how shal I forget one hoo has done sow much to instruct me in the path of virtue. My daily prayer to God is that he may keep us all in the path of virtue and piety. My daughter, Elizabeth Fulton, would be very glad to gow to that country to see you, but circumstances will not admit of it at this time. Thomas Mitchel sais he will bring my bed and things as he mooves next fall. If not I will come as soon as I can. I would be glad to say more to you, but the gentlemen are in a hurry, I must Desist at preasant.

"Dear Brother, I have neither room nor time to rite to you. I hope you with myself will endeavor to doo better for the time to come rite to me as soon as you can. I will Doo so too, and give my Best love to All my enquiring friends. Sow farewell.

THOS. McCLURE."

The following letter was dictated:

"INDIANOLA, IOWA, Dec. 16, 1866.

DEAR BROTHER.

Your welcome letter was received in due time and read with much interest. I was truly glad to hear you was yet among the living and enjoying such a good degree of health. Myself and family are enjoying moderate health. I am able to go about the place. I have been able to do but little for several years. I have been afflicted with a heart disease for several years and have lost my hearing to such an extent that I am not able to understand well, ordinary conversation. My wife is also failing very much; old age with its attendant infirmities has stolen upon us. I will now give you a brief history of myself and family. I first settled in Clinton County, Ind., and in a few years removed to Howard County, Ind., and lived there several years. I had several severe attacks of fever of different kinds and some of the family were sick almost all the time. I lost a great deal of stock by disease and finally became dissatisfied with the country, sold out and came to Iowa and settled where I now live, about six miles from Indianola. My oldest son, James Alexander, died twelve years ago. My oldest daughter, Elizabeth Fulton, is married and living in White County, Ind. Mary Ann, my next daughter, married Mr. J. P. West, is living near Virginia City, Nevada. Thos. Mitchel is married and living near by (our farms join). Abigal Caroline married a Mr. Owen, is living about a mile from me. Amos Harrison is married and living near me (our farms join). Sarah Jane married a Mr. Trimble, is living about a mile from me. I have two sons at home yet, Francis Asbury and Hosea Andrew; two sons died when small, John and William. I had one son, Amos Harrison, who served three years in the war and returned home safe. We sympathize with you in the loss of your sons, we know something of the feelings such a sad loss occasion. I own a good farm here and have a comfortable home where I expect to spend the remainder of my days. I came to Iowa in the year 1851; the country was then very new and thinly settled in this part, there has been quite a change since that time; the country is still improving rapaidly; in another year we will have a railroad completed to our county. Our country is very healthy and is well adapted to agriculture and stock raising.

"Nothing would afford us more pleasure than to visit you and talk with you of "by gone" days, but we are too old and feeble to attempt so long a journey. We would be glad to spend a few weeks among the hills of our native State, but cannot hope to realize the pleasure. We would be glad to have you and your wife visit us and see our new country. I suppose father is not living; I have heard nothing of him since I left Virginia. I should be glad to know something concerning his last days, when and where he died, etc. Some apology for my long silence perhaps is due you. For many years it has been negligence; sometimes thoughts of the past deterd me. I propose to forget the past and do better in future. I shall be pleased to hear from you often.

"Your affectionate Brother,
THOS. McCLURE."

The following obituary appeared at the time of his death, as also the following letter, and the obituary of his wife:

"INDIANOLA, IOWA, June 4th, 1871.

MR. JOHN McCLURE:

Dear Uncle—I avail myself of the present opportunity to write you a letter; which, but for the sad intelligence it must convey, I would take pleasure in doing.

I will first inform you of our loss, in the death of Father, which occurred on December 31st, 1870.

His health had been failing for several years, but he was able to be up most of the time until about ten months prior to his death when he was taken very sick; was confined to his bed most of the spring and summer; his health improved with the return of cool weather, until about a week before his death when he became very hoarse and was soon attacked with his old disease (heart trouble), under which he sank.

He was attended by the best medical skill the county afforded. His sufferings were intense, but he bore them without a murmur or complaint. We sent you a paper some time since, containing an obituary notice of his death, from which you will learn of his religious life, &c.

Mother's health is very poor this spring. Sister Sarah Trimble is living with us and keeping the house.

We should be glad to hear from you often. We would be pleased to have our relatives visit us, hoping that our beautiful and fertile country might tempt them to settle with us.

Mother joins me in expressing the highest regards to yourself and family. Hoping you will favor us with an early reply, I remain your

Affectionate nephew,
F. A. McCLURE."

"THOMAS McCLURE.—Brother Thomas McClure was born in Augusta County, Virginia, August 11th, 1795; emigrated to Russell County, Virginia, in 1817, where he was united in marriage to Miss Phoebe Hendrix. In 1820 he was converted and joined the M. E. Church, under the ministry of Rev. David Adams, of the Holston Conference. In 1827 his wife died; two years subsequent he married Ruth, the widow of Rev. H. E. Pendleton, and moved to Clinton County, Indiana. In 1851 he moved to Warren County, Iowa, where he lived till the time of his death, which occurred December 31st, 1870.

"Father McClure for fifty years was a consistent member of the M. E. Church, fully subscribing to her doctrines and usages. He was faithful in the discharge of all his public, social and private duties, and for thirty years sustained the office of class leader. His house was frequently used for religious worship, and was always the home of the itinerant. And though for the last few months prior to his death he was deprived of the privilege of meeting with

God's people for religious worship, on account of bodily infirmities, his trust in God remained unshaken and he bore his afflictions with Christian fortitude. A short time before his death, by his request, the sacrament of the Lord's supper was administered to him. We had a memorable time; his soul was filled with joy in the Holy Ghost, and when near the Jordan of death he was enabled to give glory to God, that he was going home to heaven.

"Thus after having been a bright and shining light in the church militant for half a century, he has been transplanted into the church triumphant; and though the church, his aged widow and children feel their loss, yet they know it is his eternal gain. M. S."

"DEATH OF A NOBLE WARREN COUNTY WOMAN.

INDIANOLA, IOWA, August 19, 1878.—Editor Register: A gloom has been cast over this community by the death of one of the oldest and most esteemed ladies of our county. Mrs. Ruth McClure, of this city, died this morning at 4 o'clock, her age being seventy-six years. She was born in Scott County, Virginia, from which place she moved to Indiana. With her husband and family she came to this county twenty-seven years ago, since which time she has remained here. She was a model woman and a noble mother. Four sons, Frank and Hosea of this city, Thomas of White Oak township, and A. H. of Missouri. They are all respected citizens, giving evidence of a wholesome and Christian raising. Mrs. McClure also leaves two worthy daughters, Mrs. H. H. Trimble, of White Oak, and Mrs. S. S. Owens, of this city. Mrs. McClure had been a faithful member of the M. E. Church for sixty years, and her last moments were those of a woman who had been devoted to her moral and Christian duties. O."

From these and the family Bible we gain the following facts:

He m. Feb'y 17, 1820, Phoebe Hendrix, b. March 27, 1798; d. Apr. 25, 1828, a sister of John M. Hendricks, who m. Mary McClure, sister of Thomas.

Four children:
(1). ELIZABETH FULTON, b. Jan. 17, 1821; m. March 5, 1843, Richard N. Boulden.
(2). JAMES ALEXANDER, b. Nov. 11, 1824, d. s. 1854.
(3). A daughter b. Nov. 11, 1824, and died Nov. 18th, unnamed.
(4). MARY ANN, b. March 23, 1826; m. Aug. 18, 1847, J. P. West.

He m. second, July 28, 1829, Mrs. Ruth Pendleton.
Eight children:
(5). ABIGAIL CAROLINE, b. Dec. 13, 1830, and m. Sept. 15, 1854, S. S. Owens. No children.
(6). WILLIAM H., b. March 15, 1833, d. i.
(7). THOMAS MITCHEL, b. Dec. 30, 1834; m. March 22, 1855, Mary Jane Latimer. Six children, viz:
  a. Judge John Thos. McClure, Roswell, N. M., b. April 3, 1856; m. Nov. 10, 1886, Clara Webster. Issue:
    (a). Gretchen, b. April 3, 1893.
    (b). Thos. Mero, b. July 21, 1895.
  b. Mary Caroline, b. Mar. 13, 1858, d. s. June 27, 1879.
  c. Amos Price, b. March 27, 1862, d. infancy.
  d. Dora Florance, b. Dec. 2, 1864; m. Oct. 13, 1887, L. H. Wilder. Son, Mark, b. April 12, 1891.
  e. Frank Homer, b. Jan. 1, 1867; m. June 26, 1895, Fanchon Clark. No. children.
  f. Ethelyn Dell, b. May 17, 18—.
(8). AMOS HARRISON, b. May 14, 1837; m. March, 1867, Martha Rogers. Seven children.
(9). FRANCIS ASBURY, named for Bishop Asbury, b. Oct. 23, 1839, d. April 23, 1908.
(10). SARAH JANE, b. Aug. 23, 1841; m. Oct. 25, 1866, H. H. Trimble. Four children.
(11). JOHN WESLEY, b. Aug. 18, 1844; d. i.
(12). HOSEA ANDREW, b. Nov. 22, 1847; m. Nov., 1873, Elizabeth Lucas. Two children.

Some additional facts relative to the life of Thomas McClure are set forth in the letters of his sisters, Elizabeth and Mary (q. v.). His unsuccessful career as a tanner reminds us of the early failures of Hon. Alexander Kelly McClure, LL.D., in the same business, and his drifting to Iowa, shortly to return to Pennsylvania.

LINE OF ANDREW McCLURE.

B. ANDREW McCLURE, the second son of James and Agnes McClure, was born in the north of Ireland about 1720, came with his father to Augusta county, where he married about 1742 Eleanor Wright.

The first mention of his name is Aug. 25, 1738, in Hume's Old Field Book, p. 57. His deed for 370 acres is dated Feb. 20, 1739, "being part of the Manor of Beverley in the county of Augusta," and cornering Joseph Teas, Daniel Monahan and James McClure, his father.

Judging from the records, he was a large and successful farmer. His name appears frequently in the court records and always in honorable connection.

The records at Richmond show that he was a soldier of the Revolution, having in 1775 served 68 days as a private in Capt. David McClure's company. His son, Andrew, afterwards Rev. Andrew McClure, of Kentucky, served 68 days in this same company as ensign.

In the following Augusta petition, headed by George Moffett and James Allen, followed by William McClure and others, we find the name of Andrew McClure occupying the thirteenth place.

"AUGUSTA Co., September 29, 1779.

"We, whose names are under written, having seen and considered the plan proposed by the House of Delegates for establishing the privilege of the several denominations of religious societies at the last session of the Assembly, do hereby declare our hearty and cheerful concurrence, with the same as being not only agreeable to our declaration of rights, but likewise may be a great means for laying a permanent foundation to maintain that liberty which we are so earnestly and so jointly contending for.

And therefore to our Legislature do we acknowledge our most hearty thanks as a small part of that tribute which we owe for that most just and candid proposition which appears to us to be very equally in favor of all the good citizens of this very extensive commonwealth.

It is likewise our humble desire and request that the General Assembly will be pleased to pass the said bill without the least alteration." (See State Library, Richmond, Virginia).

His will, recorded in Will Book 7, p. 168, was written Feb. 20, 1789, and proven July 21 the same year.

He mentions his wife, Eleanor, two sons, John and Josiah and a daughter, Elizabeth Trimble and her son, John Trimble. "To my son, Josiah, the tract of land I now live on, 265 acres." "That my beloved wife, Eleanor, be comfortably and sufficiently maintained by my son, Josiah." His sons, John and Josiah, executors, David and John McClure, witnesses.

There were nine children.

I. ESTHER, bap. June 17, 1743. Died in infancy.

II. ELIZABETH, bap. Nov. 10, 1745, mar. a Trimble. Her son, John Trimble, em. to Ky., where he died previous to 1828, leaving eight children, viz: Elizabeth, who mar. Frederick Stip, John, James, Napoleon, Harvey, Andrew, Franklin and Jane.

III. JAMES, born May 16, 1748, baptized by the Rev. John Craig, September 25, 1748, died in Augusta County Sept. 13, 1899.

He received from his father March 21, 1775, a tract of land. The deed reads "Andrew McClure, yeoman, to Jas. McClure, yeoman, 370 acres of land beginning at the corner of Joseph Teas and Daniel Monahan.

He was a soldier of the Revolution. The records at Richmond show that he served from October 22nd to November 24th, 1782, with Capt. Nathan Houston and Gen. George Rogers Clarke.

He mar. May 18, 1773, Elizabeth, dau. of John Kerr, of Augusta Co. She died 1824 testate. See W. B. 14, p. 468. They had eight children, viz:

 1. ANDREW, b. about 1775, m. Oct. 25, 1800, Mary Steele, dau. of Andrew Steele, of Augusta. They em. to

Green County, Ohio, and were living there 1829. (See D. B. I., p. 357, Jessamine Co., Ky.)

2. JOHN, b. about 1777, m. Aug. 18, 1808, Jane Steele, dau. of Andrew and Mary Steele, of Augusta, Rev. John McCue officiating. He died testate Dec. 8, 1829. (See W. B. 17, p. 312, Satunton, Va.).

Issue:

(1). JAMES, b. Aug. 27, 1816, and died in Augusta. Married (first) Martha Bush, of Louisa County, Va. Seven children. Married (second) 1859, Mrs. Elizabeth Jane Kennedy.

(2). ANDREW STEELE, b. Dec. 1, 1819, and died in Augusta Dec. 31, 1894. Like the others, a farmer. He m. Elizabeth Clasby, who died April 21, 1866.

Ten children:

    a. Mary E., b. Dec. 10, 1845, m. A. H. Cox, Augusta.

    b. John W., b. Nov. 11, 1847, m. Miss Foster, Louisa.

    c. Cyrus W., b. Mar. 19, 1849, d. May 27, 1851.

    d. Rebecca J., b. July 19, 1851, m. Wm. H. Chenowith, of Ohio. Two sons.

    e. Martha C., b. Sept. 29, 1853, m. Rufus Kindig. Six children.

    f. James B., b. June 20, 1855, m. Emma Whitesell. One son.

    g. George E., b. May 31, 1857, m. Liza Abel. One son. Farmer, near Crimora, Augusta Co.

    h. Laura A., b. Nov. 30, 1859. Died single in Ohio.

    i. Alice P., b. Jan 29, 1862, m. Charles Swagert, of Ohio.

(3). MARY JANE, b. Sept. 27, 1821, d. s. March 1, 1892.

3. MARGARET, b. about 1780; d. s. 1838. Her will is recorded in book 22, p. 221, Staunton, Va.

4. JAMES P., b. about 1782; m. Jan. 26, 1806, Elizabeth Strong, of Angusta Co., Rev. John McCue officiating. Em. to Green County, Ohio. Was living in 1829.

5. SAMUEL H., b. about 1785, m. on February 14, 1805, Barbara Fauber, Rev. John McCue officiating. Died in Augusta County at an early age, leaving one child, Elizabeth, who mar. on March 8, 1838, Jacob Shreckhise of Augusta County, Rev. Jas. Wilson officiating. From her descended Miss Martha McClure Shreckhise, a highly talented young lady, now living in Staunton, Va., a graduate of the Mary Baldwin Seminary.

6. JOSIAS, b. about 1787, and died about 1800.

7. ELEANOR, b. August 23, 1789, m. March 3, 1814, Jos. Peck, Rev. John McCue officiating. She died June 3, 1859, at the home of her nephew, Andrew Steele McClure. Her grave is shown in the old cemetary at Tinkling Spring. Her son, Henry Peck, was an honored citizen and for some years sheriff of Augusta County.

8. ELIZABETH, b. 1791, d. s. 1833. Her will is recorded at Staunton, Va.

IV. SAMUEL (twin brother to James, above) b. in Agusta County May 16, 1748, and died in Clark County, Ill., December 18, 1845. He received from his father, 1781, 306 acres of land in four tracts, which he and his wife, Jane, later conveyed to his brother John, (See D. B. 23, p. 390).

The following is found among the Draper papers in the Wisconsin Historical Society:

"Clark County, State of Illinois.

At the April term of the Circuit Court in and for the County of Clark and State of Illinois, begun and holden at Darwin on Monday, the 22d day of April, A. D. 1833, came Samuel McClure, who made the following declaration in order to obtain the benefit of the Act of Congress of 7th of June, 1832, relative to the surviving officers and soldiers of the Revolution. The said Samuel McClure, being first duly sworn in open court, doth on his oath say, that he was born in Augusta County, Virginia, in the year A. D. 1748: that he first volunteered for a period of three months under Capt. George Mathews; that they rendezvoused at Staunton in Virginia, from thence they marched to the head of Green-brier Creek, where they erected a fort (Warwick).

That in the next year he volunteered for and served for a peried of three months under Capt. William Anderson; that they marched to a place on Brier Creek called Clover Lick; that he subsequently served under Capt. Thomas Smith; that in this last named expedition they had three or four companies, and the companies under command of Col. Bowyer, who marched from Staunton to the south-east side of the Blue Ridge, where they rendezvoused a few days and then marched directly towards Richmond in Virginia, and when they arrived at Richmond were joined by other forces and all marched from there down James River towards Jamestown, under the command of Baron Steuben; that they were subsequently discharged at Richmond. That he again volunteered under one Capt. Zaccheus Johnson; that they rendezvoused on the top of Blue Ridge at Rock Fish gap.

That they were then joined by several other companies and marched towards Norfolk and continued their march for near two hundred miles, when they turned to the left towards the Rappahannock River and went to about three miles above York, Virginia; from thence to a place called Burnt Church, which was then headquarters.

That he served two tours of three months each under Capt. Johnson; that in the expedition under Capt. Smith he served three months; that he has no documentary evidence of his services and that John Caldwell is the only person known to him who could testify as to his services.

The court then put the following interrogatories to the aforenamed applicant, viz:

Qu. 1st. Where and in what year were you born?

Answer—In Augusta County, Virginia, in the year 1748.

Qu. 2nd. Have you any record of your age, and if so, where?

Answer—In my father's Bible.

Qu. 3rd. Where were you living when called into service

and where have you lived since the Revolutionary War, and where do you now reside?

Answer—I have since the war lived one year in Tennessee, seventeen years in Fayette County in the State of Kentucky; I now live in the County of Clark and State of Illinois.

Qu. 4th. How were you called into service—were you drafted or did you volunteer, or were you a substitute, and if a substitute for whom?

Answer—I was a volunteer.

Qu. 5th. State the names of some of the regular officers who were with the troops where you served, such Continental and militia regiments as you can recollect and the general circumstances of your service.

Answer—The general circumstances of my services, so far as I recollect them, have been already stated. I can recollect the names of many officers, but I cannot, after such a length of time, attempt to say whether they belonged to the regulars, or militia, or what particular regiment they belonged to. I recollect the names of Baron Steuben, Gen'l Scott, Gen'l William Campbell, Col. Bowyer, Major Willis, Major Long, Major Guy Hamilton, Capt. Johnson, Capt. Thomas Smith, Generals Lafayette and Washington.

Qu. 6th. Did you ever receive a discharge from the service, and if so by whom was it given and what has become of it?

Answer—I received discharges when my tour of duty expired. I cannot say who signed them, but I lost my discharges and all evidence of my services in the year 1784, under the following circumstances: I was moving from North Carolina to Kentucky and we, with several other families, were attacked and defeated on our way at Skagg's Creek.

Sworn to and subscribed in open court the day and year aforesaid. SAMUEL McCLURE.

And the court do hereby declare their opinion that the above named applicant was a Revolutionary soldier and served as he states.

1, John Caldwell, do herby certify that I was born in the year 1755; that I first became acquainted with Samuel McClure, who has subscribed and sworn to the above declaration in the year 1778. That in the year 1780 I was drafted as a militiaman under Capt. Long. That I served several months in the Revolutionary war against the British; that said Samuel McClure served also for some months to my certain knowledge, but I cannot at this day say how long. I knew him three or four years before we served together. I knew him afterwards in Tennessee and also since he resided in the State of Illinois. I have never heard any doubts expressed as to his services. I am seventy-eight years old and the said Samuel McClure is several years older than I am. I remember hearing of the defeat of Samuel McClure and his party on their way to Kentucky while I was at Richmond, in Virginia.

Subscribed and sworn to in open court the day and year aforesaid.   JOHN CALDWELL.

State of Illinois, } Sct.
  Clark County,   }

This day personally appeared before the undersigned, a Justice of the Peace in and for the County of Clark and State of Illinois, Samuel McClure who, being first duly sworn according to law, deposeth and saith that, by reason of old age and the consequent loss of memory, he cannot swear positively as to the precise length of his services, but, according to the best of his recollection, he served not less than the period mentioned below and in the following grades, to-wit: I served as a private in the year seventeen hundred and seventy-four, a period of three months and sixteen days under Capt. George Mathews against the Indians, six nations.

And I served a period of three months in the next year (1775) against the Indians under Captain Wm. Anderson. And I also served three months as a mounted Rifleman against the British under Captain Thomas Smith.

I also served two tours of three months each under Capt. Johnson against the British, when we drove Lord Cornwallis into Little York in Virginia. This last named tour under Captain Johnson, was the last I rendered the United States, and I was discharged by Captain Johnson in August in the year seventeen hundred and eighty-one, and Cornwallis was taken the same fall after I left the army. It was in June, 1774, that I entered the service under Captain Mathews. I cannot say positively, but I think that in the tour under Capt. Johnson, Abraham Smith was our Colonel and one Guy Hamilton was our Major, but in the previous tour Wm. Bowyer was our Colonel. I think one —— Long was our Major, and I think he belonged to the Regular Army. Under Capt. Smith, Charles Baskin was our Ensign. I was eighty-five years of age on the 16th day of May A. D. 1833.

I have never directly or indirectly received one cent of or from the United States or any one of the United States for my services, and I do hereby relinquish every claim whatever to a pension or annuity except the present, and I declare that my name is not on the pension role of any agency in any State. And for the services hereinbefore mentioned I claim a pension.

SAMUEL McCLURE.

June 21st, 1833."

The following obituary appeared in The Union, City of Washington, Jan. 13, 1846:

"And still another of the choice spirits of '76 has gone to his rest. Samuel McClure was born in Augusta County, Virginia, 16th May, 1748, and died at the residence of his grandson, Samuel McClure, Esq., in Clark County, Ill., the 18th December, 1845. He was a soldier of the Revolution, a brave and good man. Shortly after the close of the war he removed with his family to Kentucky. On his way they were overtaken by a party of Indians, his wife taken prisoner and his four childen butchered. He made his escape, obtained help, overtook and severely punished the Indians and secured his wife. They arrived in Lexington without property, without children, but thanks to the strong arms and stout hearts of such men as Samuel

McClure, not without a country. In early life he joined the Presbyterian Church, and his after life showed the sincerity of his vows. He voted for Washington, Jefferson, Madison, Jackson, VanBuren and Polk for the Presidency. He was a true patriot; all the ends he aimed at were his country's and his God's. He died in peace, and his departure to the spirit land more resembled the visitings of a calm sleep than the presence of the king of terrors. He was a man of good education and beloved by all who knew him. It is deemed right that when such a man passes off the stage of action his memory should receive this passing notice, though a fuller sketch of his eventful life would be gratifying to his friends."

An account of the Indian encounter at Skagg's Creek given in Collins' History of Kentucky, Vol. 2, p. 760, is as follows:

"In the year 1785 the camp of an emigrant named McClure was assaulted in the night by Indians near the head of Skagg's Creek in Lincoln County and six whites killed and scalped. Mrs. McClure ran into the woods with her four children and could have made her escape with three if she had abandoned the fourth; this, an infant in her arms cried and gave the Indians notice where she was. The Indians killed the three older children, compelled Mrs. McClure to mount an unbroken horse and accompany them home.

She was rescued the next day by her husband and Capt. Wm. Whitley."

Mr. Charles Kurtz, of Paris, Ill., under date of Nov. 20, 1911, writes:

"Samuel McClure was a very prominent citizen of Clark County. He acted as sheriff there for several years and owned a small amount of land in Edgar County."

V. JOHN McCLURE, son of Andrew McClure and Eleanor Wright, was born in Augusta Co. about 1750. Is mentioned in his father's will, 1789, and administered on his estate. He was also a Revolutionary soldier, serving as shown by Virginia records extant, in Capt. Robt. Barnel's

Company, Col. John Logan's Regiment May 3, 1781, to May 25, 1781.

Also in Lincoln Militia 1782, John McClure, Lieut. to Capt. John Boyle, Col. Benjamin Logan.

He received by patent from the Commonwealth of Virginia, Dec. 24, 1787, 1,000 acres of land in Jessamine Co., Ky. He moved to Kentucky soon after this and died single in Jessamine Co. in 1819. In Minute Book D., pp. 107-110, Jessamine Co., we find the following: "On May 17, 1819, a writing, purporting to be the last will of John McClure deceased, was produced in court by John Trimble (whose mother is a sister of McClure and one of the devisees and executors named in the will) who moved the court for probate thereof and admission to record, which was opposed by the heirs of said McClure, to-wit: David Wilson, Andrew McClure, William Sullivan and Alex. McClure."

This estate was later owned by his nephew, Andrew McClure.

VI. JOSIAH MCCLURE, born about 1752, one of the executors of his father's estate. He and his wife, Sally, were living in St. Louis Co., Mo., in 1825.

VII. ANDREW, (Rev. Andrew McClure, of Kentucky), born in Augusta Co. in 1755; graduated from the Augusta Academy (now Washington and Lee University) with A. B.

After leaving school he did some work as a surveyor in Augusta. "May 22, 1773, surveyed for Mr. Geo. Gibson the annexed plot of 1,264 acres of land in Augusta County, Beverely Manor lying on a branch of Christian's Creek called Black's Run.

AND. McCLURE."

This was copied by Mr. Edward Frazer, of Lexington, Ky., from "Andrew McClure, his Book of Mathematics," in Mr. Frazer's possession. There is a court record where Andrew McClure on Aug. 17, 1773, was appointed a road surveyor for Augusta Co. He, like his father and brothers, was a Revolutionary soldier. The Virginia State Library gives the fragmentary record that he served 20 days in 1775 with Lieut. Francis McClure and 88 days as Ensign under Capt. David McClure.

We find further in the House Journal, on February 12, 1777, the following: "Congress having resolved that this Board forthwith cause to be levied two hundred men to garrison forts Pitt and Randolph, to be commanded by two captains, four lieutenants and two ensigns. In conformity thereto Robt. Campbell and John Robinson are appointed captains, Thadly Kelly and Andrew McClure first lieutenants." The record shows that Capt. Robinson, 1st Lieut. Andrew McClure, 2nd Lieut. Brenton and Ensign Wallace were placed in command at Fort Randolph, known later as Point Pleasant.

He later entered the ministry of the Presbyterian Church, being received as a condidate by Hanover Presbytery at the Augusta (Stone) Church, November, 1781.

"He visited Kentucky in 1784, but returned to Virginia and was ordained pastor of Roan Oak. He could not forget, however, the charms of Kentucky, and in 1786 removed thither with his family. In 1787 he organized the Salem and Paris Churches, and in 1789 took charge of the latter, where he remained till his decease in 1793 in the 39th year of his age."—Davidson's History of the Presbyterian Church in Kentucky, p. 83. See also Bishop, p. 282, and Bishop's Rice, p. 69.

In Collins' history of Kentucky, vol. I, p. 461, "Rev. Andrew McClure removed to Kentucky in company with Mr. Craighead." He was licensed at New Providence Oct. 24, 1782. His parts of trial were, Popular Lecture, 2 Cor., 3:18; Exegesis, Rev., 1:10 to end; Sermon, John, 17:17. The summer following he preached at Wilson's, in the Sinks, at Indian Creek, Rich Creek, Anthony's Creek, Jackson's River and the Pastures.

He was ordained at Bethel May 19, 1784. In vol. II, written minutes of Hanover Presbytery, there was "a call from Sullivan County, N. C., for Mr. Andrew McClure in particular, or any other whom Presbytery might appoint. A call also from the north and south forks of Roan-Oak to

Mr. Adrew McClure. It was ordered that both calls be presented to Mr. McClure. He accepted the one from Roan-Oak. The congregation of the north and south forks of Roan-Oak having signified their earnest desire to have Mr. McClure ordained as soon as conveniency would admit, Rev., 22:4 was assigned him for his trial sermon."

"The first church in Bourbon County was Presbyterian, organized in 1787 at Paris by Rev. Andrew McClure, who had been preaching in the place occasionally for three years."—Collins' Vol. II, p. 72.

He was present, a charter member, and the first clerk of Transylvania Presbytery, the first Presbytery of Kentucky, organized in the court-house at Danville, Ky., Tuesday, Oct. 17, 1786. See Collins', Vol. II, p. 457, and Minutes of Transylvania, Vol. I, p. 96.

In the miuutes of the meeting of October 1, 1793, held at Cane Run Church, we find, "The Rev. Andrew McClure departed this life Aug. 25, 1793, in the 12th year of his ministry and the 39th of his age."

He was buried in Paris, Ky.; his grave unmarked.

In a family Bible, now owned by Mr. Edward Frazer, Lexington, Ky., we find the following:

"Commencement of matrimony August 29th, 1782,
   between And. McClure and Rebeckah Allen.
James Allen McClure was born Nov. 20th, 1783.
Eleanor Wright McClure was born November
   25th, 1785.
Polly McClure was born January 5th, 1788.
Andrew McClure was born Sept. 5th, 1790."

His wife, Rebeccah Allen, was a daughter of James and Mary Allen, of Augusta Co., and is mentioned in his will written April 28, 1788. The other children were Margaret Bell, Agnes Shields, Elizabeth McNair, Mary, Francis and James Allen, Jr.

In Dr. Jos. A. Waddell's Address at the 150th Anniversary of Augusta Stone Church Oct. 18, 1899, we read:

"One of the first elders was Captain James Allen, who lived near the place called Willow Spout. The earliest mention I have seen of him was in 1742, when he was appointed constable by the court of Orange County for the part of the country west of the Blue Ridge, an immense domain thinly settled, except immediately around Staunton. In 1756 he was captain of a militia company of sixty-eight men. Next we find his name signed in Oct., 1776, to a paper forwarded by several companies of militia and free holders of Augusta to the first Legislature of the new State on the subject of religious liberty. The signers demanded that all religions denominations within the Dominion be forthwith put in full possession of equal liberty without preference or pre-eminence. Up to that time no dissenting minister was authorized by law to perform the marriage ceremony, and all the people were required to contribute to the support of the Established Church."

Dr. Waddell further says" that in Oct., 1783 or 1784, a large party of people went from this congregation to Kentucky,—Trimbles, Allens, Moffetts and others. They went on horseback by a circuitous route, and very dangerous on account of hostile Indians. Every man carried a rifle and every woman a pistol, and they did not fail to take their Bible, the Confession of Faith, the Shorter Catechism and Rouse's Version of the Psalms." It was probably at this time and with these people that Rev. Andrew McClure first went to Kentucky.

His will is recorded at Paris, Ky.

"IN THE NAME OF GOD, AMEN.

I, Andrew McClure, of the County of Bourbon, and State of Kentucky, being very sick and weak in body, but of perfect mind and memory, thanks be unto God; calling unto mind the mortality of my body, and knowing that it is appointed for all men once to die, do make and ordain this my last Will and Testament; that is to say principally and first of all I give and recommend my soul into the hands of Almighty God that gave it, and my body I recommend to the earth, to be buried in decent, Christian burial at the discretion of Executors, nothing doubting but at the general resurrection I shall receive the same again by the mighty power of God. And as touching such worldly estate, wherewith it has pleased God to bless me in this life, I give, demise and dispose of the same in the following manner and forever: I give and bequeath to Rebekah, my beloved wife, my plantation whereon I now live, with all the profits and advantages accruing therefrom by cultivation to her and her children during her life, and at my wife's death I give and bequeath to my two little daughters, Eleanor Wright McClure and Polly McClure, the said plantation to be equally divided between them, and to be freely possessed and enjoyed by them and their heirs forever.

Also I give and bequeath to my two little sons, James Allen and Andrew, my plantation contained in two deeds and lying on Howard's Creek in Clark County, to be equally divided between them in quantity and quality, and to be freely possessed by them and their heirs forever, together with all the advantages accruing therefrom from now and henceforward. Also my two negroes, Isam and Doll, I grant and allow to be set at liberty, if from the judgment of their owners and the Executors, they shall have served truly and faithfully; Isam until the age of thirty-five years and Doll thirty-three.

Also my cash and cash bonds I give and bequeath to my two sons, James Allen and Andrew, to be speedily put to interest and kept for the use of their education.

Also my negro girl, Doll, I give and bequeath to my dearly beloved wife, Rebekah, during her servitude; likewise my negro boy, Isam, as long as she remains single, but after marriage I allow the boy to be hired during his servitude and his wages applied for the use of the education of my children.

Also I allow ten pounds of the arrearages due to me on Howard's Creek, when it can be collected, to be put into the hands of the Transylvania Presbytery for Charitable purposes.

* * * * * * * * * *

In witness whereof I have hereunto set my hand and seal this fifteenth day of August, one thousand seven hundred and ninety-three.

ANDRED McCLURE, (S. S.).

Teste: William Maxwell, William Craig, Alex. Martin."

Four children.

1. JAMES ALLEN McCLURE, 1st., b. in Augusta Co., Va., Nov. 20, 1783; went to Kentucky with his parents in 1786; thence to Illinois in 1833, settling on a farm near Carlinville. The last years of his life were spent in Washington, D. C., in the employ of the government, where he died July 28, 1849. He is buried in the Congressional Cemetery, where a stone marks his grave. He married, about 1806, Frances Dickerson, dau. of Martin and Rebecca Dickerson, of Jessamine County, Ky. She died in Illinois March 13, 1844.

The following letters, furnished by Mr. Edward Frazer, of Lexington, are of interest:

"WASHINGTON CITY, D. C., March 17, 1846.

DEAR SISTER.

I promised you when I left your house last that I would write pretty soon to you after my arrival at this place. Well, as an excuse for not writing sooner I have no very good reason except laziness to give. You know well that I have all my life been sorely afflicted with that cause. I now write you and thank God that my health and strength and unprofitable life are preserved, and that I am in the land of the living and in the state of repentance. I arrived here safely on the last of November and swore in, and entered upon the duties of my office on the 2nd of December last, and have never missed a day in the discharge of my official duties. I am on a salary of a $1000 per annum, paid monthly. Expenses in this city are high. Board cannot be had in decent families for less than $4 per week, besides your washing and other expenses. I am boarding with Col. Laughlin, the second officer (Recorder) in the General Land Office. He is a Tennesseean with whom I had a slight acquaintance in that State. He is a very learned man; has large library of books and newspapers from every part of the Union. He is wealthy and does not take boarders, but took a fancy to me and invited me to come and live with him. I have a fine room to myself and all my wants attended to. I never did live in a more agreeable family in my life, and it is close to our office. I have never as yet been to hear any of the debates in Congress. It is a mile from our office to the Capitol. We are closely confined from 9 in the morning till 3 in the evening without anything to eat till about 4 o'clock, which is our dinner hour. I generally take a biscuit in my pocket and eat it at about 12 o'clock with water, which refreshes me very much. I do not know how such close confinement may agree with me when the warm weather sets in. Old Uncle Samuel McClure died on the 15th Dec'r last, at a very advanced age; upwards of 90. There was a short obituary notice of his death in the papers here. With the light that brief notice gave and what I knew of the old man myself, I, with the aid of a friend, have written out a more extended obituary and sent on the manuscript this day to the editor of the Kentucky Gazette at Lexington for publication in that paper, and I have requested the editor to send a paper containing the notice to the post-office at Lexington addressed to you as a relative of the deceased, and also a paper addressed to my Brother Andrew at Nicholasville. I have made mention in the above notice, of the names of our venerable father and grandfather, who have long since been numbered among the dead. I want you by all means to see what we have written.

I heard from home the other day; all my dear children were well then. Has Amanda written to the girls? She promised me pointedly she would. Have the boys send on the salve for William. Write or let William or the girls or James write. I should be glad to hear from you, for next to my own children you stand highest in my affections. I have no news but what you have in the newspapers. There will be no war. Give my respects to all the

children without exception. Excuse errors, for I am writing by candle-light and cannot see the lines and have not time to make corrections. I am my dear sister,
Most affectionately your brother,
JAMES A. McCLURE.
Mrs. Polly Sullivan, Lexington, Lafayette Co., Ky.
I will write to-morrow night to Brother Andrew."

"WASHINGTON CITY, August 7th, 1849.
MY DEAR SIR.

Being fully sensible of the painful shock which the contents of my last letter would necessarily inflict on the already lacerated feelings of Mrs. White and yourself, I have delayed all further communication on the melancholy subject until now, hoping that time in the interim, in some small degree at least, would so much assuage and mitigate our common sorrow, as to enable me to write more calmly, and you and the faithful sharer of all your joys and sorrows to receive the farther relations which I feel it my duty to make concerning the death of your beloved kinsman, in a spirit of greater resignation than you could have commanded while the unexpected event was so fresh, and overwhelming in its natural effects on the mind of affectionate children. I say children, because Mr. McClure loved *you* in all the fullness of affection of a father for a son. No man ever loved more ardently, sincerely and impartially than he did every member of his family. Without objects on whom he could constantly lavish the affections of a heart made up of love and universal charity, the world would have been to him a dreary wilderness, and life itself a meaningless blank. But to those who knew him, as you have known him, I need not say a word as to the goodness and purity of his heart, or of his kind and affectionate disposition.

My chief object in writing, is to inform you and his children, as truly as I can, of a few details relating to his last days and hours which I was incapable of setting down in any sort of order when I wrote to you last. Although Mr.

McClure hoped, as did his friends, until the last fatal change in his case, that he would be able by easy and careful travel to get home—at all events to get to Cincinnati—which he esteemed as home; but for many months he had been fully satisfied that his days were numbered and that his life, if he got up and out in the world again, would be but short. He had thought and reflected so maturely on the subject that he had become reconciled to death and willing to die whenever God in His good prividence should choose to call him hence. As to death, his only wish and constant prayer was that he might be enabled to die, whenever the time should come, without bodily pain and long suffering. As to religion, he talked with a number of friends—clergymen of different denominations, Presbyterians, Baptists and Methodists. They often prayed with and for him. He said he felt comfort from such visits and exercises. He consulted his Bible much in the course of the last year. He has left it to be sent home, a small, beautiful pocket Bible, which he used most. He could read it conveniently lying in bed, and had it nearly always in reach. While he could sit up he used a Bible of mine, in royal octavo form, on account of the large, plain print, it being an edition of the Bible society, and of course without notes or commentaries. I suppose he has declared to me, in the fullest confience, fifty times in the last six months, that he had full and perfect trust in the mercy and goodness of God, and that he confided and trusted himself, as to all his hopes of this world or eternity, to the mercy and justice of his Creator. He said again and again, that that gracious providence which had preserved him from so many evils through life, would assuredly bring him in safety to the end, and to the better and more perfect state of another and happier life, in which he now fully believed. On Sunday, the day before he died, he declared to myself and others his perfect resignation to the will of God in whatever was to happen, and in being secure in that happy immortality in which his late pious father and beloved wife had trusted and believed. He often adverted to a promise he had made to his wife on her death bed to endeavor to become a Christian, and how anxious he had ever been to fulfill it. In our private intercourse, long before he was smitten down by disease, he had told me of this promise, and often dwelt upon his determination to fulfill it, with tears. Early in his confinement, in the latter part of last year, and frequently since, he informed me how much he had regretted through life, that when he was a youth in Kentucky, and after quitting school, he had read Pain's attacks on religion with great avidity, and many other similar works, and that his guardian, the late Judge Trimble, of Kentucky, was an avowed infidel and never discouraged him from such reading. Though he never became an infidel himself—never adopted deistical opinions—yet, he said his course of reading, and the unbounded faith he had in Trimble's judgment, made him ever ready to doubt, and ever ready to quote the opinions he had read, and such as he had heard Trimble advance in conversation against the truth of the Christian religion. Trimble was truly a man of splendid talents, and possessing an amiable and excellent character, and being a man universally esteemed, made his opinions more dangerous to young men. He said that reading the works of Paley, Dick and Lord Brougham on natural religion, in Illinois, had enabled him more nearly than anything else, to shake off those early imbibed prejudices. During the summer and fall of 1846, and winter of 1847, when he lived in my house with me, I bought and persuaded him to read with me, and talk over together, Bishop Butler's Analogy between natural and revealed religion. This work he confessed put to shame all the infidel books he had ever read. Still they had left the early impression of youth which had to be rooted out, but I had no doubt of his being an earnest seeker after the real truth during the last two years of his life, and I have no doubt of his having found it to his own satisfaction. It was in this that he was enabled to put away from him the very great fear of death which haunted him so much when I first met and lived in his company here.

As to his worldly matters he made no *Will*. He always said his children knew precisely what disposition he wished made of his effects. As to some small things I have heard him speak, Brande's Encyclopedia of Science, valuable large book—and a copy of Webster's large Dictionary with Worcester additions, publishes by the Harpers. I think he wished Milton to have them if he perseveres in his design of becoming a scholar or professional man. His little Bible I have no doubt he desired some one of his daughters to have, but I never heard him say.

Being removed from office some weeks before his death I had all his sympathies in my favor. I have no doubt of his having been more pained at the event mentally than I was myself. We talked freely about it. As I had purchased a cottage and few acres of ground in the suburbs in a shady, high, dry and well watered place, surrounded by groves and wholly out of the noise of the city, though in its chartered limits, we had agreed that as soon as he could ride a little better than he found he could then on trial, and as a means of gaining strength by change of air and scenery, and as a training for his journey until he got strength enough, he would go with me where we would live in the quiet of country life, with as much of books and friends as we might desire to see or read; and finally, that if travel should turn out to be impracticable till frost or next spring, or if he could not get off at all, that we would remain together in our retreat, doing the best we could—living in frugal comfort and in charity with all the world, for neither of us since we resided here had ever had the slightest dispute or controversy with any man. Our friendship, which commenced in a slight acquaintance in Tennessee, about thirty years ago, had increased daily, as I sincerely believe, for the whole term of our service in Washington. In our retreat, sick or well, we would have been mutual helps. We had proceeded in the plan so far as to have made estimates of expenses, supplies, and every outfit necessary to begin with. Although I ardently desired to see him get off home, and had urged him to go home last October and November when he got up and worked five or six weeks in his room; a time when Judge Young had promised him, through me, leave of absence, and when no change of administration was expected. His fears then principally, were that if he got to Cincinnati, or Illinois, by easy stages of travel, he might become sick there, be compelled to overstay his time, and that possibly he might get so decreped from his stricture as not to be able to come back at all, though able for the business of his desk if here, and that in some of these contingencies he might lose his place. I had no such fears—at any rate of the last, for Judge Young has continued until lately and he and myself since Judge Young resigned, and before I was removed got his pay continued to 1st of September next if he lived so long, and leave of absence till then. This could have been done if he had been lying sick at home under the circumstances of his case. The few weeks he was able to write last fall, I am sure is the only time within which time he could have gone home since he was taken first with serious illness about 4th July, 1848.

When he lost his sons he suffered greatly on both occasions. On one of the occasions—the last—I feared for his reason for some days. I got him to travel with me, when we went to Baltimore, thence down the bay in a steamer to Norfolk, Virginia, where we inspected the public works and national shipping, especially the Pennsylvania, the largest ship in the world. Thence we went up James river 130 miles to Richmond, and examined the paintings and statues of Washington and Lafayette in the capitol. Thence home. In this journey of about 500 miles, all in cars and steamboats, we were engaged nearly three weeks. It was during that excursion that I discovered the wonderful effect change of place, scenery, company, &c., had upon him, both bodily and mentally; and as it was so very favorable, I have ever during his whole last illness endeavored to get him to go home—to ride out when he could—and lastly, it was the good effects of that trip which made us both think that to retire with me to my hill in the suburbs, when we could not take any longer trips, would be salutary. In every design we have attempted for his recovery and assistance, Col. Cabell, the two Youngs, and other friends have all concurred and all in agreement with the advice of his excellent physician, Dr. Hall. I find that the proof of the immortality of the soul, and life in eternity after death, being known to David, as is disclosed in the incident of his determination to go to his dead son who could not come to him, as related in 12th chapter of 2 Samuel, is a portion of the Bible which Mr. McClure has been in the habit of reading and meditating on deeply. In my Bible he has marked with pencil in the margin from the 15th to the 23rd verse, inclusive. The blessed book being true I now hope he is with his sons, his sainted wife and pious father and others who proceeded him. In meantime may heaven bless you and yours.

S. H. LAUGHLIN.

To Wm. McKim White."

There were fifteen children.

(1). MELINDA McCLURE, born about 1808 and married Oct. 5, 1829, Larkin B. Smith in Kentucky.

(2). AMANDA McCLURE, b. about 1810, was married to William McKim White in Carlinville, Illinois, Aug. 2nd, 1836, by. F. M. S. Smith. They removed to Cincinnati, Ohio. Several children were born, only two living to maturity, Antoinette and Frank. She died in Cincinnati in July, 1883, having survived her entire family.

(3). ANDREW McCLURE, b. about 1812, was married to Mary Fishback August 17th, 1841, by F. M. S. Smith. They had one child, Grundy McClure.

(4). Mary McClure, b. about 1814, d. s.

(5). ELIZABETH JANE McCLURE, was born Jan. 12th, 1817. She was married to Dr. Grundy Blackburn July 28, 1835, by S. M. Otwell, and died 1908. They had five children.

a. Adelaide, b. Oct. 10, 1836, d. Oct. 20, 1852.

b. Edmonia, born 1838, d. Jan. 1852.

c. Grizzell, b. Sept. 11, 1841; married Grove Lawrence, a native of Connecticut. Three children:
  (a). Grove P. Lawrence, living in St. Louis.
  (b). Hal Lawrence, married, lives in Seattle, Washington. Three sons.
  (c). Lida Blackburn Lawrence married Charles Wiley, of Charleston, Illinois. Later they moved to Seattle, Washington. They were both drowned while on a pleasure trip in British Columbia, July 13th, 1910. They left three children:
    Lawrence Wiley, born Sept. 27, 1895.
    Marian Wiley, born Sept. 27, 1897.
    John Wiley, born 1905.
    All living in Seattle, Washington.
d. Newton B., b. 1845, d. i.
e. Emmons, b. May 1. 1849, d. s.

(6). JOHN McCLURE, born September 22, 1818, and died February 16, 1894. Married Elizabeth Campbell, daughter of William and Ann Campbell, April 6th, 1858. There were three children:
  a. W. C. McClure, born Feb. 28, 1859.
  b. John McClure, born Feb. 11, 1860, died October 11, 1905.
  c. Theodore McClure, born January 7, 1865, and died August 3rd, 1911.

(7). JAMES ALLEN McCLURE, 2nd, was born in Shelbyville, Tennessee, April 12th, 1820. In 1832 he moved with his parents to Carrollton, and thence to Carlinville, Ill., where he was engaged in agricultural pursuits. He died March 2nd, 1901. He married Ellen Collins, daughter of Enos and Margaret Groniger Collins, on December 20th, 1855. Her family came from Pennsylvania to Ohio, thence to Illinois. She was born in Ohio November 21st, 1833, and died March 28th, 1903, in Carlinville, Ill. They had five children namely:
  a. Col. Charles McClure, b. Sept. 28, 1856, d. Nov. 19, 1913. The following, which appeared in the *Car-*

linville Democrat at the time of his death, gives the outline of his life:

"COL. CHARLES MCCLURE.

Early Thursday morning, November 20, relatives in this city were informed by cable of the death on the preceding day of Col. Charles McClure, at Fort William H. Seward, near Haines, Alaska.

On Saturday previous they had been informed by wire that Col. McClure had undergone a very serious operation for an illness that began November second, and that there were but slight hopes for his recovery. On Sunday morning a reassuring message was received and in the absence of information to the contrary during the succeeding days, his relatives and friends here had hoped that the crisis had passed. Accordingly, the death message came as a great shock.

Charles McClure was born on a farm three and one-half miles south-east of Carlinville, on September 28, 1856. He attended the district school and later was a student at Blackburn College. Subsequently he won in a competitive examination and became a cadet at the United States Military Academy at West Point. From that institution he graduated in June, 1879. By choice and by education he was a soldier. He had an exalted idea of his profession and quickly demonstrated his soldierly efficiency. He gave his country the very best of his ability, and in whatever capacity he served, from Second Lieutenant to Colonel, he did his best. Early in his military careex he studied law, and was admitted to the bar in Illinois in August, 1885. The government, at various times and in various ways, availed itself of his broad legal knowledge.

Col. McClure was united in marriage in this city on October 3, 1882, with Miss Mae Walker, daughter of former Senator and Mrs. C. A. Walker. One child was born of this union, Charles W. McClure, now a first lieutenant in the regular army and with his company doing service on the Mexican border.

COL. CHARLES McCLURE,
1856-1913.

The remains will be brought from Alaska to this country and interred in the Arlington cemetery at Washington, D. C. The uncertainty of ship transportation at this season of the year renders it impossible to determine at this time when the funeral services will be held.

Col. McClure served his country thirty-eight years. From the beginning to the end he was faithful and earnest. He was modest and gentle, but forceful and aggressive, when force and aggression were demanded. He was a student, a thinker; he believed in preparation in all things. He was a natural soldier. He sought to understand his duties and endeavored to do well his part. As a young officer on the frontier in 1879; as the instructor in military tactics at the University of Illinois for three years; as an officer at posts in all parts of this country; as the Judge-Advocate of the Department of Columbia for four years; as special assistant to the Judge-Advocate General in the revision and codification of the military laws of this country; as the special counsel for the Government in the prosecution of Captain Carter; as a soldier on the battlefield in both Cuba and Philippines in the Spanish-American War; as Assistant to Gen. Ainsworth, Chief of Staff of the U. S. Army in Washington; as Colonel of the 30th U. S. Infantry at San Francisco, and finally as the commandant of the military posts in the territory of Alaska, he was always earnest, patriotic, efficient.

Though stricken down in the prime of life, with his work only partially finished, Col. McClure has left a record which is a credit and an honor to the county and the State of his birth."

He was buried in Arlington Dec. 8th, 1913, where the officers and men of his regiment, as a token of love and esteem, erected a monument to his memory.

Author, "The Opinions of the Judge-Advocate," a standard work.

His son, Lieut. Charles Walker McClure, of the 7th U. S. Infantry, was born July 23rd, 1883, married April 17,

1909, Justine Semmes Moran, daughter of John Valle Moran and Emma Eltridge Moran, of Detroit, Mich.

b. Frank McClure, born Sept. 28th, 1856. He married Adelle King, daughter of Captain Lucien King and Almira Lemen King, of Kane, Illinois, Sept. 9th, 1879. He is Vice-President of the South Arkansas Lumber Co., with headquarters in St. Louis, Mo. Three children have been born to them:

 (a). Charles King McClure, born Sept. 12, 1881, in Kane, Illinois; married Mary Hicks, daughter of L. B. and Clara Lambert Hicks, of Henderson, Ky., Oct. 8th, 1908. They have two children, Charles King McClure, Jr., born July 10th, 1910, and Lambert Hicks McClure, born November 22, 1912.

 (b). Florence Adelle McClure, born April 11th, 1884, in Kane, Illinois.

 (c). Sudie Louise McClure, born June 19th, 1887, Kane, Illinois.

c. Milton McClure, Jr., born September 3rd, 1858. Died April 9th, 1913. He married Rose Orwig, daughter of William and Jane Orwig, in Beardstown, Ill., May 20th, 1885. He achieved great success in his chosen profession, the law.

He was the nominee of the Republican party in 1909 for Justice of the Supreme Court, Fourth Illinois District.

Appropriate memorial exercises were observed by the Cass County Bar Association on May 19, 1913, with addresses by Hon. J. Joseph Cooke, Hon. Guy R. Williams, Hon. Chas. A. E. Martin, Hon. A. A. Leeper and Judge Harry Higbee.

He left one son (a) Floyd Milton McClure, b. March 31, 1890, in Beardstown, Ill.; m. March 1914, Beulah Glendinning, dau. of Frank and Alice Glendinning, of Beardstown. Lawyer.

d. James Enos McClure, born August 8th, 1867. On September 30th, 1897, he was married to Emma Florence Parker, daughter of Henry and Harriet King Parker, of

Kane, Illinois. He was admitted to the bar in 1893, but preferring newspaper work is now editor of The Carlinville Democrat. He has held many positions of trust in the State. Two daughters:

 (*a*). Harriett McClure, born Nov. 9th, 1900.

 (*b*). Dorothy McClure, born Sept. ——

 e. Edmonia Blackburn McClure, born March 21st, 1870. Married August 8th, 1894, Jesse Peebles, son of Judge L. P. and Sarah Odell Peebles, of Carlinville, Illinois. A distinguished lawyer. They have three children:

 (*a*). Martha Allen Peebles, b. June 3rd, 1898.

 (*b*). Don McClure Peebles, b. Sept. 7th, 1900.

 (*c*). Pauline Peebles, b. Feb. 3rd, 1904.

(8). AMELIA McCLURE, b. ——, died July 19th, 1851, of cholera. Married Alexander McKim Dubois, Oct. 31st, 1844, at Carlinville, Ill. Their children were:

a. Nicholas Dubois, born April 7th, 1846, married Orlena Eliza Daws, of Carlinville, Ill., November 16, 1871, and now lives in Springfield, Illinois. Their son, Alexander Daws Dubois, was born in Springfield, Illinois, Dec. 19, 1875. He married ——, and has a daughter, Charlotte Amelia Dubois, born Springfield, Ill., August 3, 1897.

b. Catharine McKim Dubois, born May 7, 1849, and married Ethan Allen Snively at Carlinville, Ill., Feb. 23, 1876. Mr. Snively has held many prominent political positions in Illinois. They now live in Springfield, Illinois.

c. James McClure Dubois, born June 15, 1851, died July 17, 1851.

(9 and 10). FELIX AND PAUL McCLURE, twins. Felix died Nov. 20, 1847, in Cincinnati, Ohio, where he was in a medical school.

Paul died in Carlinville of cholera.

(11). MARTHA McCLURE, born 1822, died Nov., 1887. Married Dr. Levi Woods, of Carlinville, Illinois, on March 15, 1844, by F. A. Ramsay. Two children were born of this marriage:

 a. Frances Woods, b. Feb. 9, 1848, married Judge Whitlock, of Jacksonville, Ill., on October 19, 1869, died May 1, 1892.

b. William McKim Woods, born June 28, 1850. Was a physician. Married Lolah Walker, daughter of Senator C. A. Walker and Permelia Dick Walker, on May 23rd, 1882. One son was born to them, namely, Chas. Herbert Woods, born March 4, 1883. Is an attorney in Carlinville, Ill. Married Nov. 26, 1910, Norma Hoblit, daughter of A. Lincoln Hoblit and Josie Stanley Hoblit, of Carlinville, Ill. They have one son, Dick Hoblit Woods, born on March 20, 1913.

(12 and 13). TULLY AND REBECCA McCLURE, died in infancy.

(14). FRANCES McCLURE, born on Jan. 27, 1831, at Nicholasville, Jessamine County, Ky. Died Nov. 9, 1888, at Carlinville, Illinois. Married Samuel Sayword Gilbert, an attorney of Carlinville, August 6, 1851. Children:

a. Amelia, born August 22, 1852; died Sept. 3, 1852.

b. Edward Addison, an attorney at York, Nebraska, born July 6, 1854, at Carlinville, Ill. Married Louise Mayo, daughter of Samuel T. and Elizabeth Palmer Mayo, of Carlinville, January 1st, 1878, in Carlinville. They have the following children:

    (*a*). Elizabeth M. Gilbert, now Mrs. L. W. Childs, York, Nebraska.

    (*b*). Frances Louise Gilbert, now Mrs. E. G. Brown, Long Beach, California. They have two living children, Gilbert J. Brown, born April 6, 1907, and —— Brown, born Oct. 30, 1913. One child, Frederick J. Brown, died in infancy.

    (*c*). Edward Addison Gilbert; was married to Mina Alexander, of York, Nebraska. They have one child, Edward A., born in 1911.

    (*d*). Margaret Palmer Gilbert, now Mrs. Fisher, of York, Nebraska, has one child born in December, 1913.

c. Samuel S., Jr., b. Dec. 14, 1856, d. Aug. 9, 1858.

d, Claribel, b. Nov. 14, 1859, d. March 14, 1860.

e. Charles Frederick Gilbert, an attorney at York, was born August 12, 1862, at Carlinville. Married Pearl Barcefer, of Kansas City, Mo., December 16, 1891. No children.

f. William White Gilbert, dealer in real estate at Muskogee, Oklahoma, was born January 27, 1865, at Carlinville, Illinois. Married Mary Bronaugh, daughter of Perry Bronaugh, of Virden, Illinois, in June, 1903.

(15). MILTON McCLURE, SR., born in Nicholasville, Ky., in 1832. Died January, 1903. Married Martha K. Neale, of Springfield, Ill., in 1854. They had two children:

 a. James Allen McClure, born August 7, 1856. He devotes his time to real estate and other interests. In 1905 he married Mary Agnes Davis, daughter of Mr. Landon and Myra Davis, of Bolivar, Missouri.

 b. Harriett B. McClure, married Thos. B. Mellersh, of Cincinnati, Ohio. She was born Feb. 1858, and died in San Francisco, Cal. They have two children, Neale Mellersh, now of San Francisco, Cal., and Claude M. Mellersh, of San Francisco, California.

2. ELEANOR WRIGHT McCLURE, second child of Rev. Andrew McClure, born in Augusta Co., Va., Nov. 25, 1785, was taken as an infant to Bourbon County, Ky., 1786. She was killed July, 1817, by lightning striking the Presbyterian Church where she was worshipping. There is extant a printed notice of her funeral from her home on Hill street, Lexington, Ky., July 21, 1817, at 10 o'clock A. M.

She was twice married. First to John Lawson, of Lexington, Ky., and second to L. McCullough.

Her only child, MARY PIERCE LAWSON, was born June 16, 1807; m. March 23, 1825, John Bowman and died Sept. 28, 1862. Her thirteen children:

 (1). ELEANOR LAWSON, b. Dec. 30, 1825, and d. in Gatesville, Texas, in 1907 in her 82nd year. She m. Aug. 3, 1843, Philip S. Woodward and had five children, three of whom married.

 (2). ISAAC BOWMAN, b. March 1, 1827, m. Grandison

B. Smith, Dec. 21, 1848; nine children, four of whom married.

(4). JOHN BOWMAN, b. Feb. 1, 1830, d. i.
(5). LAWSON, b. Jan. 3, 1835, d. i.
(6). AMELIA SULLIVAN, b. April 19, 1837, m. Oct. 7, 1856, Guilford H. Slaughter; died Dec. 25, 1909. Four children, a. Mary Henry, m. B. Whitfield; no children. b. Sallie Amelia, m. Edmund Leyon, five children. c. May Byrd, m. G. W. Meriweather, three children. d. John H. G.; two adopted children.
(7). GEORGE W., b. Jan. 7, 1840, d. s. Jan. 1, 1862.
(8). MARGARET CAMPBELL, b. Sept. 20, 1841, d. i.
(9). JOSEPH LAWSON, b. Dec. 20, 1842, d. i.
(10). JOHN, b. Dec. 13, 1844, d. i.
(11). ELIZABETH, b. May 17, 1846, d. i.
(12). SUSAN ANN, b. Dec. 8, 1847, m. John K. Smith; no children.

(13). SARAH McCLURE, b. Sept. 18, 1852; m. Feb. 23, 1881, Joseph H. Jones; no children. She lives at St. Bethlehem, Tenn., the only one of the thirteen children living, and very kindly furnished the above information.

The following letter, furnished by Mr. Edward Frazer and published with his permission, is from Eleanor McClure's second husband. It is written in a neat, clear hand:

"COLUMBUS, Feb. 6, 1837.

DEAR FRIEND WILLIAM.

As the bearer, Mr. Mason, intends being in Lexington, I imbrace the oppertunity to write you a few hasty lines, just to state we are all in our usual health. I have not heard anything from yourself or family since I had the pleasure of seeing you last faul. When my son Davie was here a few weeks past he informed me that it was his impression that your honored and very aged father had departed this life, but at what time he was unable to say.

It was my impression when I last saw him that his days were but few, but that he was ripened for an incorruptable crown. I should be very glad indeed to have a letter from you.

I had a letter lately from Hugh M. Allen stating the death of Martha Allen, which took place on the 16th ulto. As he was prevented from seeing her on her death bed by reason of confinement from a severe fall by which two of his ribs were broken, could not give us the particulars of her death, but was informed Jane intended to write us on that subject shortly.

I have had several letters from son Harvey since my return from the West. His own and wife's health, together with greater part of the children, was still bad, very bad. His object is to move back to Columbus as soon as he can arrange his affairs, that is some time the insuing spring, although he is at a difficulty what he will engage in as a way of living. I believe I told you of Matilda's son, who was in ill health when I was in Illinois. In a recent letter she states he grew worse and worse until he lost use of all his limbs, also the use of speech, which remained for upwards of six weeks, but has happily recovered his usual health and use of his speech and limbs, &c.

My own health was much benefited by the late journey, and thus far much better than it was the last winter and summer. Part of my hours are occupied in teaching Daughter Gravis to make vests and overalls, otherwise I should be called an entire gentleman (that is nothing to do). Robt. Milton has quit farming and now lives in Cincinnati and engaged in some business as a clerk I believe. I find two Kentucky gentlemen were on business with our Legislature (about slavery), but nothing I believe has yet been transacted respecting their mission unless their memorial to the Honorable body. After all, negroes will run away from your State, be harboured and forwarded on their way to Canada, and you cannot help yourselves. If the whites do not engage in it, there are sufficient number of blacks in the State to do so, of which they boast without scruple. In fact, altho a large majority of our citizens are not Abolitionists, there appears an appathy about the matter, so what is every bodies business is no bodies at all.

You may see that petitions have lately been presented our Legislature from sundry colored people about the benefit of our school system, &c.

The Dutch have had their advocates in the house to have appropriations of the school fund so that the German language may be kept up; also a petition has been before the Legislature for like privilege to the Welch, to keep up the native language, &c. Our internal improvement and school system are inviting a debt on the State which has raised our taxes to such a degree as to be really burdensome and the prospect of increasing the State tax alone is about 14 mills to the dollar. My tax on a 42 foot lot with common improvements, was upwards of $64, besides city tax of upwards $15, and no doubt as the last year increased more than 30 per cent. the next will be in proportion. But it is useless to complain, lest you be called an enemy to the poor.

Since Mr. Smith went to the great house at Frankfort I have no news from Lexington. As I have to write a couple other letters, I must conclude by saying I got a compleat loco-foco letter from Jas. McClure since my return. They are all well. He says there are two of his daughters living in Cincinnati. I hope to see them when I pass through that place at a future time. Our love to Sister Sullivan and family.   Your friend,
                            L. McCullough.
Capt. William Sullivan, near Lexington, Ky."

3. MARY McCLURE, third child of Rev. Andrew McClure, was born in Bourbon County, Kentucky, Jan. 5, 1788, and died Aug., 1880. She married Capt. William Sullivan (Feb. 23, 1780—July 11, 1842), a son of William Sullivan, Sr., who moved from Virginia to Fayette Co., Ky., sometime prior to 1779. Twelve children, all born in Fayette County:

(1). JAMES WILSON, b. Jan. 11, 1807, d. May 2, 1878.
(2). ELEANOR, b. Nov. 2, 1808, d. Oct. 9, 1816.

(3). AMELIA, b. Nov. 10, 1810, d. Dec. 6, 1831.

(4). ANDREW, b. Oct. 26, 1812, d. Oct. 12. 1816.

(5). JOHN, b. Sept. 16, 1814, d. Aug. 1, 1815.

(6). MARY McCLURE, b. May 7, 1816, d. Dec. 9, 1836; m. —— Hewitt; son, Wm. Andrew, b. Aug. 13, 1836, d. i.

(7). ELIZABETH, b. Oct. 5, 1818, d. Jan. 12, 1889.

(8). WILLIAM M., b. Nov. 27, 1820, d. June 13, 1857, m. Dec. 17, 1850, Lucy B. Allensworth, No children.

(9). HARRY, b. April 8, 1823, d. i.

(10). MARTHA, b. Aug. 18, 1825, d. s. Feb. 15, 1901.

(11). ALEXANDER CAMPBELL, b. April 22, 1829, d. i.

(12). ADISON, b. April 25, 1831, d. i.

JAMES WILSON SULLIVAN, the oldest of these children, was married three times; first on July 16, 1839, to Jane S. Gatewood, who died Sept. 18, 1842, leaving no children.

He married second, May 19, 1846, Maria Louisa Fleming (July 19, 1817—July 20, 1862), a daughter of Leonard Israel and Nancy Marshall Fleming, and granddaugher of Col. Wm. Fleming, of Virginia. (For a sketch of Colonel Fleming see Whitsett's Biography of Caleb Wallace, Waddell's Annals Augusta County, and Foote's Sketches of Virginia 2nd, p. 268).

Her great grandmother was Mary Marshall McClanahan, aunt of Chief Justice Marshall, and wife of Rev. Wm. McClanahan, of Virginia and North Carolina.

They had six children, viz:

a. Andrew McClure, born in Lexington, Ky., March 11, 1847, graduated at University of Kentucky (Transylvania), 1869, and settled at St. Louis, Mo.; married July 25, 1885, Jessie Peel Young. No children. He is the senior member of the law firm of Sullivan and Wallace, St. Louis.

b. James Richard, b. May 3, 1853, d. ——; m. Sallie Hamilton, of Lexington, Kentucky; two children, John, d. i. and Annie, wife of Rev. Joseph G.

Armister, pastor of the Christian Church at Walla Walla, Washington. One child.

 c. Bryan, b. Jan. 15, 1855, d. s. at Lexington, Ky.
 d. Fleming, b. Jan. 15, 1855, living single in St. Louis.
 e. Mary, b. July 10, 1857; m. Dec. 17, 1884, Wm. F. Stanhope, of Fayette County, Ky., and died without children May 15, 1885.
 f. Nannie, b. Aug. 12, 1858, d. s. at Lexington, Ky.

They are all buried in the family lot in the Lexington, Ky., Cemetery.

He married third, Feb. 14, 1867, Sarah Boone, a widow from Bourbor County, Ky. No children.

ELIZABETH SULLIVAN, the other child of Mary McClure, who left descendants, m. Joseph F. Frazer, b. Spotsylvania Co., Va., Feb. 11, 1818, and d in Kentucky Oct. 12, 1883. Four children:

a. Elizabeth, d. s.; b. Martha, d. s; c. Mary Sullivan; d. Edward.

The two latter now live in Lexington, Ky., and have rendered valuable assistance in the preparation of this work.

William Frazer, their remote ancestor, settled near Norfolk, Va., early in the 18th century, where he remained until his sons were educated at William and Mary College. Moving to Spotsylvania Co., his son, James m. Elizabeth Foster, d. of Anthony Foster of that county. Five sons and one daughter.

Anthony Frazer, the oldest of these, was b. March 22, 1754. Was first Ensign and later Lieut. in the Revoluary War. He married Hannah Herndon. Ten children, of whom Edward Frazer, the third son and fourth child, was born Feb. 1785. He married Elizabeth Frazer, d. of his uncle, John Frazer and Betty Fox. One son, Joseph F. Frazer.

4. ANDREW McCLURE, fourth child and youngest son of Rev. Andrew McClure, b. Bourbon County, Ky., Sept. 5, 1790, and died Jessamine County, Ky., Aug. 18, 1849, of cholera after an illness of twelve hours. He was for a

good many years a successful merchant of Lexington, Ky., but during the last years of his life devoted himself almost entirely to his farm. His home, a beautiful estate in Jessamine County, is still known as "The McClure Place."

He m. Jan. 23, 1827, Rachel Sarah Barton, (b. January, 1790, d. Nov., 1874), d. of John Barton, a family well known both in Virginia and Kentucky.

They had one child:

(1). SARAH BARTON McCLURE, b. June 27, 1828, d. Aug. 17, 1866. She m. Feb. 21, 1854, Isaac Shelby (Dec. 28, 1815—July 1, 1873), son of General James Shelby and grandson of Isaac Shelby, the first Governor of Kentucky.

Two children:

a. Sarah Barton, b. at Lexington, Ky., Sept. 20, 1859; m. Oct. 14, 1884, Edmund Shelby Kinkead (Oct. 14, 1856—Oct. 2, 1910), a son of Judge Wm. B. and Elizabeth Shelby Kinkead, of Covington, Ky.
Two children:
(a). Edmund Shelby Kinkead, Jr., b. Oct. 3, 1885.
(b). Elizabeth Fontain Kinkead, b. Oct. 29, 1887.

b. James, (July 24, 1861—June 23, 1862).

LINE OF ELEANOR McCLURE.

C. ELEANOR McCLURE, the third child and oldest daughter of James and Agnes McClure, was born in Ireland about 1725, and died in Augusta County, Va., in 1799. Her will is on file in box 41, office of the Clerk of the Circuit Court, Staunton, Va.

She married about 1755, Hugh McClure, who was probably a son of William McClure of page 21.

They owned a good estate near Fishersville, Augusta County, where Hugh died 1872.

They left seven children:

I. ISAAC, who died single in 1828. His will, written Feb. 5, 1828, was proven April 21, 1828, by the oaths of John McClure, Samuel H. McClure and

Elizabeth Kerr. David Bell and Hugh McClure, Jr., administrators.

II. JOHN, d. s. 1820.

III. DAVID, who on Sept. 29, 1795, married Elizabeth Holmes, d. of Samuel Holmes, of Shenandoah.

In his will, proven February, 1834, (See W. B. 19, p. 375, Staunton), he gives "the Old Mansion property, as laid off by Isaac McClure for the support of their brother, Hugh," to his son, Hugh, Jr.; the land on which he resided to his son, Samuel H. McClure; two hundred acres of the land formerly owned by his brother, John, deceased, to be laid off for the support of his son Isaac. His three daughters, Elizabeth, Rachel and Eleanor, were to be paid $700 each.

His personal property was valued at $2,332.67, and he owned bonds worth $2,180.

Of his six children mentioned above:

1. HUGH, was born Oct. 29, 1796, and died Oct. 7, 1876. A large farmer, a member of Tinkling Spring Church, where his grave is marked.

His wife was Jane Bell of Augusta County. No children.

2. SAMUEL H., died single, 1834. For his will see Book 20, p. 74.

3. ISAAC, born in 1800, died single Oct. 24, 1887. His grave is marked at Tinkling Spring.

4. ELIZABETH H., married Samuel Bell Kerr, of Augusta. She died about 1877, leaving five children.

(1). Alexander Kerr, who married Fannie Homan and left five children:

    a. Charles H. Kerr, m. Annie Borden.

    b. A daughter, m. M. A. Coiner.

    c. Frank A. Kerr, m. Sadie Borden.

    d. Howard L. Kerr.

    e. Richard Kerr, m. Jessie McNeil, of Staunton, Va.

(2). David McClure Kerr, m. Kate Myers and left five children, viz., Arthur B., Wilbur M., Emmett W., Lee and Elvan.

(3). Elizabeth E., m. P. J. Link, died 1900. Two children:
   a. Alice V., m. C. C. Thompson.
   b. Hester N., m. W. H. Landes.

(4). James T. Kerr, m. Sarah Myers. Three children, viz., Hugh McClure, David Bell and J. Newton.

(5). Samuel Holmes Kerr, m. Mary E. Bondurant. Four children:
     a. Hugh Holmes Kerr, Commonwealth's Attorney, Staunton, Va.; m. Sarah E. Rock.
        (a). Elizabeth Holmes.
     b. E. Bondurant Kerr, m. (2nd) Lucy G. Waddell.
     c. Janetta Waddell.
     d. Elizabeth Barry.

5. RACHEL, m. J. B. McCutchan, of Augusta Co.

6. ELEANOR, b. Oct. 9, 1804, and died single Dec. 6, 1891. Her grave is marked at Tinkling Spring, new cemetery.

IV. AGNES McCLURE, m, Robert Pilson, and was living in Kentucky in 1808. Her children, Hugh, Anna, Richard and Polly, were living in Ohio in 1833.

V. ESTHER McCLURE, m. July 7, 1806, Isaac Trotter, Rev. John McCue, pastor of Tinkling Spring, officiating.

VI. HUGH, "of unsound mind" died single.

VII. JOEL, died single.

LINE OF JANE McCLURE.

D. JANE McCLURE, the second daughter and fourth child of James and Agnes McClure, was born in Ireland; came with her parents to Augusta County 1738, and married 1757, (second wife) Capt. Archibald Alexander, born Cunningham Manor, Ireland, February 4, 1708. They lived near Timber Ridge, Rockbridge Co., in which church he was a ruling elder.

The Alexander Genealogy has been carefully worked out in Roger's Memorials of the Earl of Sterling and the House of Alexander, and in a chart by Mr. Francis Thomas Anderson Junkin, LL. D., of Chicago.

The facts of his life are so fully given in Waddell's Annals of Augusta Co., aud in Alexander's Life of Dr. Archibald Alexander, p. 6-7, that it is not necessary to repeat them here.

His will, recorded at Lexington, Va., was written Nov. 29, 1779, and proven Feb. 1, 1780.

There were seven children by this marriage:

I. MARY, b. July 4, 1760. She m. first, John Trimble, b. Aug. 24, 1749, and d. 1783, leaving one son, James, b. July 5, 1781; and died near Nashville, Tenn., 1824. For further information see Waddell's Annals of Augusta. She m. second, Lewis Jordan and moved to Tennessee.

II. MARGARET, b. Feb. 1, 1762, d. s.

III. JOHN, b. July 28, 1764, and d. 1828. He was a soldier of the Revolution, substituting for his half brother, William, who was married and had a family. (See Alexander's Life of Dr. Archibald Alexander, p. 15).

IV. JAMES, b. Oct. 4, 1766, m. Martha Telford.

V. SAMUEL, b. Feb. 1769, m. McCoskie.

VI. ARCHIBALD, b. March 3, 1771, m. Isabel Patton.

VII. JANE, b. 1773, m. Rev. John W. Doak, of Washington Co., Tenn., son of Rev. Samuel Doak, D. D., (1748-1830). She was the mother of Rev. Alexander A. Doak, of Washington Co., Tenn., father of (1) Rev. A. S. Doak, b. Sept. 10, 1846, now pastor of the Presbyterian Church Lenoir, Tenn., (2) Rev. S. H. Doak, father of W. H. Doak, M. D., the father of Rev. A. H. Doak, pastor Presbyterian Church, Wilmore, Ky.

E. JAMES McCLURE, JR., the third son and fifth child of James and Agnes McClure, was born in Ireland about 1730; was the youngest of the children that came with his parents to Augusta County. He received pay 1758, for militia service. See Hening, Vol. VII, p. 181.

He administered on his father's estate in 1761, receiving according to the terms of the will, one half of the farm.

The only further information, is that found in Chalkley, vol. III, p. 384: "Aug. 27, 1761, Jas. McClure, of Craven County, South Carolina, to John Ramsey, 408 acres devised to James and his brother Samuel, by will of father, James McClure, and descended to grantor by survivorship, being the same that James the father bought of William Beverly, 6th June, 1739. Delivered: Wm. Ramsey 5th Oct., 1772."

It is probable that he emigrated to Craven County, N. C., instead of Craven District, S. C. The records at New Bern, Craven Co., N. C., give a deed to James McClure 1760. There is no further mention of his name. Under Wills, the Craven County records contain that of Jacob McClure, 1837, who mentions a son, Elisha McClure.

F. SAMUEL McCLURE, the fourth son and sixth child of James and Agnes McClure, was born after parents came to Virginia. He was baptized by Rev. John Craig Nov. 9. 1740. According to the terms of his father's will, 1756, he was to share the farm equally with his brother James. It seems from the record of his brother James that he died prior to 1761.

G. ESTHER McCLURE, the youngest child of James and Agnes McClure, was born about 1741; was baptized by Rev. John Craig Nov. 8, 1741; is mentioned in her father's will 1756. No further record.

With these we conclude the records of the descendants of James and Agnes McClure, the original settlers.

OTHER McCLURES IN AUGUSTA COUNTY.

A. FINLEY McCLURE, possibly a brother of JAMES, settled in Augusta County as early as 1739. The name suggests his connection with the Finley family, well known both in Augusta Co. and in the north of Ireland. His deed for 440 acres "A part of the mannor of Beverly" joining Patrick Campbell and David Mitchel, is dated Feb. 28, 1739, Orange C. H. His name spelled McClewer, appears 1742 on the muster roll of Capt. John Christian's company of militia, third in a list of seventy-six. The Augusta records give two transfers of land, viz., 1742 and 1748, when he seems to have located in the western part of the county. The last mention of his name is 1768, when he witnessed a deed for John Tate.

He was probably the father of MICHAEL McCLURE, who lived in that neighborhood, born about 1750 and married between 1774 and 1778 Mary Wetzell. (Chalkley, Vol. I, p. 377).

He witnessed the will, 5th Nov., 1780, of Archibald Gilkeson. Is mentioned in Wm. McPheters' list of tithables, Augusta, 1781.

We find him in the Rockingham County census of 1784, giving his family as six.

Having married a Wetzell, we are inclined to the opinion that he is the ancestor of John McClure, of Franklin, Pendleton Co., W. Va.

Martin Wetzell, of Augusta County, in his will written Feb. 9, 1795, and proven April 8, 1795, names John McClure, his grandson, his sole heir and executor. (Will Book I, p. 20, Chalkley III, p. 249). This Wetzell family mentioned in McClung's Western Adventure, was well known. Martin's brother, Lewis Wetzell, was the most renowned Indian fighter of his time. They were sons of John Wetzell, an early settler.

The following information was furnished me by Mr. John McClure, Franklin, W. Virginia:

"My grandfather, JOHN McCLURE, was born in Augusta County 1777, and died in Franklin, Pendleton Co. W. Va., 1858. He was related to the Wetzells, but I do not know just how the relationship comes in. He came to this county in 1798 and married Elizabeth McCoy Oct. 15, 1799. Two children:

1. ELIZABETH, who died young.
2. JOHN, born 1805 and died 1854. He m. Feb. 25, 1829, Sidney Judy (Dec. 5, 1808—March 19, 1888). Five children:
    (1). Elizabeth, b. 1829, d. 1912. Married Amby Harper.
    (2). Katherine, b. 1833, d. 1857. Married Jacob Harper.
    (3). John, b. June 1, 1838, and m. 1867 Rebecca J. Skidmore.
    (4). William, b. 1846, a Confederate soldier. Killed at Lynchburg, Va., June 17, 1864, in Hunter's raid.
    (5). A child, died in infancy."

John McClure, now (1914), seventy-six years old, is one of the best known citizens of his section of the State. As president of The Farmers' Bank of Pendleton, a large land owner and dealer in live stock, he has for years been looked upon as a leading figure in the material development of Pendleton and adjoining counties. Both he and his wife are highly esteemed members of the Franklin Presbyterian church.

He may be a descendant of John McClure, who was doubtless a kinsman of Finley and James, and who lived near what is now Dayton, Rockingham County, Va.

B. JOHN and MARY McCLURE came to Augusta county as early as 1740 and like JAMES were members of Rev. John Craig's congregation. His Baptismal Register gives the names of five of their children, viz:

James McClure, baptized Nov. 2, 1740.
Ann McClure, baptized March 29, 1741.
Mary McClure, baptized Nov. 28, 1742.

Jean McClure, baptized July 14, 1745.
Elizabeth McClure, baptized May 21, 1746.
Thomas McClure, baptized Sept. 5, 1748.

They moved, 1749, to the south fork of the North River of the Shenandoah, within the present limits of Rockingham County. His deed for 400 acres from George II is dated Dec. 15, 1749. (See records, Richmond, Va.)

His name is signed to the South Fork road petition of 1749, along with Daniel Smith, William Logan and others. (Chalkley, vol. I, p. 433.) Also with Daniel Harrison, James Magill, Gabriel Pickens, et al., 1754, to a petition to "The Worshipful Court of Augusta County." (Chalkley I. p. 313.) He bought, 1751, 387 acres from Daniel Harrison on Muddy Creek, on north side of North River, deeding the same date his original 400 acres to Silas Hart. His farm joined James Magill.

Another transaction is dated 1768, witnessed by Benj. Logan and Elizabeth McClure. Again in 1767, when John and Mary McClure convey to John Houston 200 acres on Muddy Creek. (Chalkley, vol. III, p. 481.)

Was sued by James Magill, 1770, for saying "Would hang as high as Gilderoy."

The last mention of his name in the Augusta records is 1773. As his home after 1778 was in Rockingham, any facts after that date would be in the records of that county, which unfortunately were burned.

Of his children—

I. JAMES McCLURE was b. about 1739. Possibly the James McClure of Cherokee County, S. C. (See McCurdy's S. C.)

II. ANN, b. 1741, m. John Logan, Augusta County. Emigrated to Kentucky; ancestors of the Hon. John G. Carlisle.

III. MARY, b. 1742, m. August 8, 1763, Col. Benjamin Harrison of Rockingham County, son of Daniel Harrison. Qualified Lieutenant-Colonel May 8, 1778. She died 1815, leaving sixteen children, viz:

1. Robert, m. 1784, Polly Harrison.

2. Daniel, m. 1784, Anne Patton
3. John, m. 1792, Ann Tallman.
4. Benjamin, m. March 22, 1791, Polly, d. of John Hall, by Rev. Wm. Wilson. Em. to Kentucky.
5. James, m. Anna Wilson. Em. to Kentucky.
6. Edith, m. Samuel McWilliams. Em. to Kentucky.
7. Margaret, m. 1797, Ezekiel Logan.
8. Jane, m. Rev. Wm. Cravens.
9. Peachy (1777-1848), m. 1802, Mary Jane Stuart.
10. Fielding, m. 1800, Ann Quinn. Em. to Kentucky.
11. William, m. (1) Jane Young, (2) —— McClure. Em. to Kentucky.
12. Jesse, m. 1794, Elizabeth Wilson.
13. Thomas.
14. Parthenia, m. 1804, Reuben Harrison. Em. to Kentucky.
15. Marellah.
16. Endocia.

(See also Waddell's Annals, p. 152.)

Among the many living descendants of Mary McClure is Miss Jannetta Burlingham of Shullsburg, Wis.

IV. JEAN, b. 1745, m. Col. John Logan, of Augusta County, cousin of the one above. Em. to Kentucky. (See Green's Historic Families of Kentucky). Daughter Mary m. Otho Holland Beatty, whose daughter Jane m. Joseph Ballenger, of Lincoln County, Ky.

V. ELIZABETH, b. 1746, m. after 1768, a McKey of Rockingham County, son, Joseph McClure McKey.

VI. THOMAS, b. 1748. Was doubtless the Thomas McClure of N. C., Ensign, Revolutionary War, enlisted April 1, 1776. Mentioned in Wheeler's History of N. C., p. 80, from Salisbury District. Was wounded in the fight at Hanging Rock August 6, 1780, in which battle Capt. John McClure was mortally wounded. In my judgement he is the great grandfather of Mr. Adolphus B. McClure, of Aztec, New Mexico, who, in a letter dated July 18, 1911, gave me the following:

"My family originated in Scotland, from whence a Presbyterian minister and his two sons emigrated to Ireland. There is a tradition that they were weavers by trade. Coming to America they settled in Virginia, where my grandfather, Thomas McClure, was born about 1775. He had a brother, John, an Indian fighter, who went to Kentucky. Also a sister, Matilda, who married a Skidmore and went to Kentucky. Her grandson, Peyton Skidmore, died in Aztec, N. M., in 1902, and left an only daughter, Mrs. Lucy Hoyle, one-eigth Cherokee Indian. She is now a widow, wealthy and beautiful. Thomas McClure moved to N. C. prior to 1800, and died near Dalton, Ga., 1865.

His children, born in N. C., were (1) Andrew, (2) John, b. 1800 and died at the foot of Pike's Peak in 1880. He left a son, William, who now lives near Gordon, Erath County, Texas; (3) William, who lived and died near Waynesville, Haywood County, N. C., leaving two sons, Pinkney and Rowland; (4) Thomas; (5) a daughter, who m. a Moody and had two sons,—Joseph who lived in Jackson County, N. C., and Rev. Hiram Moody, a Baptist preacher, who was in Texas 1875 and afterwards in Oklahoma; (6) Col. Joseph McClure, born near Raleigh, N. C., July 16, 1810, m. October 29, 1835, on a farm that is the present site of Birmingham, Ala., Patience McLain; commanded a regiment in the Mexican War, and died in Fayette County, Ala., Nov. 2, 1855. He left five children:

a. Martha A., b. Murray County, Ga., Jan. 23, 1837, m. J. J. Spain April 23, 1855, and now lives with her family near Alto, Texas.

b. Thomas M., b. Murray County, Ga., Sept. 14, 1840, lives near Alto, Texas; is the father of Rev. J. T. McClure, of Dallas, Texas.

c. Columbus C., b. Murray County, Ga., August 9, 18—, died in Winchester, Va., February 9, 1862. A marble slab shows his grave in the N. E. corner of the old cemetery there. A Confederate soldier.

d. Adolphus B., b. Fayette County, Ala., February 25, 1848, now lives at Aztec, N. M.; interested in real estate.

e. Henry C., b. Fayette County, Ala., June 2, 1850, and died in Palo Pinto County, Texas, October 9, 1889.

As further proof of my conclusion, a Skidmore family lived in Rockingham County, neighbors to the McClures. (See McClures in N. C.)

Green, in "Historic Families," states that ROBERT and WILLIAM McClure of Kentucky were brothers of Ann and Jane Logan. If so, they were younger brothers, born after 1748.

VI. ROBERT McCLURE is mentioned in Augusta County records (Chalkley, vol. III, p. 144) as a member of Capt. John Gilmore's company in his expedition against the Cherokees, 1778.

The Revolutionary War records at Washington, D. C., show that he served as Sergeant in a Virginia infantry regiment. Collins' History of Kentucky, vol. II, p. 554, speaks of him as being in Lincoln County, Ky., 1784, and the Richmond, Va., records show that he received, Feb. 1, 1795, by virtue of two treasury warrants, 29,300 acres in Harrison County, Ky. Is mentioned as being in Ohio County, Ky., June 9, 1796.

VII. WILLIAM McCLURE was a soldier of the Revolution, as shown by the Washington, D. C., and Virginia records. Emigrated to Kentucky, settling at Stanford, Lincoln County, in 1789. Through the interest of his personal friend, Gen. Benj. Logan, moved to Shelby County. (See Collins, vol. II, p. 474.) Wife, Rebecca. Among their children were:

1. JANE ALLEN McCLURE, b. Sept. 3, 1783, was living June, 1871. She married a Stuart; parents of the late Judge James Stuart, of Owensboro, Ky.

2. ROBERT McCLURE. No record.

C. PATRICK McCLURE, mentioned once in Chalkley, vol. III, p. 25. "June 17, 1752, Martha Mahan's bond as administratrix of Patrick McClure, with sureties Patrick Martin, Wm. McFeeters." Probably the father of Patrick McClure, a Revolutionary soldier from Virginia. Martha Mahan was doubtless his wife.

D. WILLIAM MCCLURE, who witnessed the will of James McClure, 1756, and died prior to 1761. No record of either deed or will. He was doubtless the father of,

I. HUGH MCCLURE, of p. 123.

II. JOSIAS MCCLURE. In 1797 Josias McClure acquired a tract of land, "being the same granted to James McClure by patent." He mentions in his will, written 1814 and proved 1817, his wife, Jane, and his friends, Isaac, David and Hugh McClure, Ann Hutcheson, wife of George H., Sr., Agnes Pilson, d. of Hugh McClure, deceased; Mary McKenny, d. of And. McClure, deceased; John McClure, son of Andrew and grandson of Mitchel; and Hugh McClure of unsound mind. Jane McClure his wife was a daughter of William Johnson, Augusta County. For her will, 1817, see W. B. 12, p. 389.

III. WILLIAM MCCLURE. He was doubtless the signer of the Augusta Petition on p. 89. Chalkley, vol. III, p. 473, gives his deed for land on Middle River, in Beverley Manor, August 16, 1768. Teste: John Stuart, Hugh Allen, Andrew McClure. April 16, 1793, William and Elizabeth McClure sold 198 acres on Middle River. In 1818 William McClure, in Richmond, Va., sold for $3,500, one-third of 500 acres in Augusta County. There are McClures now living in Richmond, probably his descendants.

McCLURES IN ROCKBRIDGE COUNTY.

This county being set off in 1777 from Augusta and Bottetourt, information prior to that date is found in the records of those counties. There is nothing to show that the family is closely related to that of Augusta. As it is morally certain that both families came from near Raphoe, County Donegal, it is probable that the Augusta family is descended from John McClure and the Rockbridge family from Arthur McClure, both of whom were Ruling Elders in the Raphoe congregation, 1700. They were doubtless related.

Of the Rockbridge family,

HALBERT MCCLURE came to the county about 1740. As there is no record of his importation it is probable that he first settled in Pennsylvania. The earliest mention of his name is 1742, on the list of Capt. McDowell's militia company. His deed is recorded in Book I, p. 203, dated March 19, 1746. "Benjamin Borden to Halbert McClure, 230 acres on North branch of James River, corner of Samuel McClure." This was in the bounds of Timber Ridge congregation, where, in 1753, he signed the call for Rev. John Brown.

His will, proven 1754, is recorded at Staunton, Va. (See Chalkley, vol. III, p. 34. "Halbert McClure, gentleman." He mentions his wife Agnes, a nephew, Halbert, son of his brother John, deceased, and two sons, Alexander and Nathaniel.

A. ALEXANDER MCCLURE, b. about 1717, was a member of Capt. John McDowell's company, 1742. His deed for land on Mill Creek is dated 1747. Was a Ruling Elder in Timber Ridge Church, signing the call to Rev. John Brown, 1753, and represented his congregation in Hanover Presbytery, 1760. His wife Martha was probably a daughter of James and Martha Moore, of Timber Ridge. He died 1789. His will is recorded at Lexington, Va. Eight childrsen:

I. HALBERT, b. about 1748; was living in Rockbridge, 1793.

II. JAMES, b. about 1750; probably the James McClure who was a Revolutionary soldier, private on Capt. Elliot's ship. The Richmond, Va., records give his deed 1772, from George III, for land in that part of Botetourt that in 1777 became a part of Rockbridge County.

III. NATHANIEL, b. about 1752. Deed from George III for 150 acres on James River, Botetourt County, 1774. Probably emigrated to Kentucky. The Richmond, Va., records give a grant to Nathan McClure, 1786, in Lincoln County, Ky.

We find in Collins' History of Kentucky, vol. II, p. 685, for 1788: "Lieut. Nathan McClure following Indian horse thieves was shot and mortally wounded, dying the succeeding night in a cave. He was an active officer and his loss was deeply deplored."

IV. ALEXANDER, JR., settled in Botetourt, now Rockbridge County. His wife Agnes was probably a first cousin; daughter of Moses McClure. He died about 1810. Her dower was set off Nov. 25, 1828. Five children, viz: John who died about 1817, Thomas, Moses, Isabella who married Andrew Hall, Catherine, who married James Taylor. (See Rockbridge County records.)

V. SAMUEL, a Revolutiouary soldier; member of Capt. Thomas Rowland's company, Col. Wm. Fleming's regiment, 1777, Botetourt County. He married, January 24, 1782, Rosanna Steele, daughter of Nathaniel and Rosanna Steele. A son, Nathaniel, mentioned in the latter's will, 1795.

VI. JOHN, who, by the terms of his father's will, received land in Kentucky. Possibly the John McClure of Harrison county, Ky., pentioned 1817. Probably the John McClure of Botetourt, in Capt. John Murry's company, Col. Wm. Fleming's regiment, 1775. (See Dunmore's Wars).

VII. SUSANNA, No record.

VIII. MARTHA. No record.

B. HANNAH MCCLURE, d. of Halbert and Agnes McClure, m. Robert Allison, of Rockbridge County. Eight children: James, Mary who m. Davidson, Agnes, Robt., Francis, Halbert and Janet. Robert Allison died 1769, Alexander and John McCluer, executors. Chalkley, Vol. III, p. 109.

C. MOSES MCCLURE was probably the oldest son of Halbert and Agnes, born about 1710 and died 1778 intestate. Was a member of Capt. John McDowell's company, 1742. His large farm on the south side of North River cornered Nathaniel McClure, John McClure and Thomas Paxton. Was a member of Timber Ridge Church 1754. He married, about 1745, Isabella Steele, daughter of David Steele. She died 1797. Her will is recorded at Lexington, Va. Their children were four sons and five daughters, viz:

I. DAVID MCCLURE, Captain in the Revolutionary War

and on March 4, 1777, was appointed Lieutenant-Colonel of the county of Ohio, Ky. (See Richmond, Va., records.) He married Eleanor Steele, daughter of Nathaniel Steele, and is mentioned by him in his will, 1795. In 1805 he sold his large farm to his brothers, Alexander, Moses and Halbert, and emigrated to Grant County, Ky. The records give the names of two sons:

1. HALBERT McCLURE, mentioned in the will of Nathaniel Steele, 1795.

2. DAVID McCLURE, JR., mentioned in the Rockbridge records, 1817.

One of these is doubtless the ancestor of the late Moses McClure, who died an old man in Grant County, Ky., 1907, and of Ezra K. McClure, the son of John McClure, of Crittenden, Grant County, Ky.

II. HALBERT McCLURE, born about 1750 and died in Rockbridge County about 1830. He married a daughter of Nathaniel and Rosanna Steele, and is mentioned in their wills. Son,

1. MOSES McCLURE, born about 1785 and died May 10, 1829. He married, about 1812, Elizabeth Jones. Six children:

 (1). Alexander, born September 22, 1813.

 (2). Nicholas J., born Novemember 23, 1815.

 (3). Mary Steele, born August 26, 1817.

 (4). Moses F., born June 6, 1819.

 (6). David K., born December 27, 1827.

 (5). William Preston McCluer, born April 12, 1822, married, May 11, 1843, Nancy Jane Shields. Six children.

  a. Napoleon Bonaparte, born March 30, 1844, and died June 4, 1904. Married Sallie Wilson April 9, 1784. Four children.

   (a). Harry Scott McCluer, born March 28, 1875. Married Nora Echols September 18, 1906. Two children, Lois Argyle, born June 2, 1907, and Elizabeth, born Jan. 13, 1908.

   (b). Frank Wilson McCluer, born October 13, 1876.

Married Daisy Lee Butler Sept. 9, 1906. Dentist, Lexington, Va. Two children:
 a. Anna Lee, born Oct. 14, 1907.
 b. Frank Wilson, Jr., born Feb. 13, 1909.
(c). Annie F., Sept. 16, 1878—Aug. 16, 1886.
(d). E. Blanche, Oct. 23, 1884—July 28, 1886.
b. Bettie, born April 20, 1848, d. s.
c. Emma J., born May 13, 1850.
d. Rachel P., born November 7, 1853.
f. Mattie S., born June 20, 1858.
e. John W. born August 25, 1856. Farmer and merchant, Fairfield, Va. Six children.
 (a). William A. born June 25, 1895.
 (b). Margaret E., born May 10, 1898.
 (c). Mary Ethel, April 28, 1901—Dec. 9, 1903.
 (d). Eleanor Blanche, born July 16, 1902.
 (e). John Donald, born March 15, 1905.
 (f). Walton Malcolm, born Sept. 20, 1907.

III. MOSES McCLURE, mentioned with Alexander and Halbert McClure in Rockbridge records 1807. In 1832 Moses McClure bought from Alexander H. McClure and his wife Jane of Union, Ky., their land on north branch of James River.

IV. ALEXANDER H. McCLURE was probably the youngest of the children, b. Oct. 31, 1774, and died in Kentucky May 9, 1843. He m. first, MARTHA ELLIOT, of Rockbridge County on Oct. 29, 1795. Five children: James, Hannah, Susan, Peggy, and Patsy, who m. Jacob Myers, parent of Alexander Myers.

He m. second, Sept. 20, 1810, JANE GIBSON, and soon after emigrated to Grant County, Ky. Seven children:

1. THOMAS, who m. a Coons; thirteen children.
2. ELIZABETH, who m. Harry Brown; ten children.
3. ALEXANDER, d. s.
4. NANCY; m. a McClure from Ohio.
5. JOHN, b. Sept. 20, 1820, d. Nov. 5, 1871. He married Ann Berthena Larvell. Five children:
 a. Rev. James W. McClure, a Presbyterian minister,

living in Cythiana, Ky. He m. in S. C. a Miss Steele, a descendant of Archibald Alexander Steele, of Va. Five children.

 b. Robert R.
 c. Moiver J.
 d. Mary E.
 e. Lawrence W.; d. when three years old.

 6. WILLIAM HARVEY MCCLURE, m. Lucinda Brown. Three children:
  a. Archibald Alexander McClure, m. a Ransom.
  b. Thomas Wesley McClure, m. a Linsey.
  c. Betty Alice McClure, d. at sixteen.

 7. KITTY MCCLURE, d. s. at twenty-two.

Of the five daughters of Moses and Isabella McClure, AGNES married her cousin, Alexander McClure; Rosanna m. a Love; SUSANNA, died single; ISABELLA married David Steele; BETSY probably married Alexander McClure, son of William.

D. NATHANIEL MCCLURE, son of Halbert and Agnes McClure, born about 1712. Wife, Mary. Probably the same that settled first on Middle River of the Shenandoah, member of Captain John Christian's militia company 1742, and whose son, Alexander, was baptized by Rev. John Craig, March 10, 1749. In Rockbridge County, was a member of Capt. John Buchanan's militia company 1742. His farm (deeded 1747), was on Mill Creek, cor. Moses McClure. Was constable 1745; member Timber Ridge church 1753, signing the call for Rev. John Brown. He died 1760 and his wife MARY died 1767. For their wills see Chalkley, Vol. III, pp. 62 and 101. Ten children:

I. HALBERT, b. about 1740. M. Mary Henderson and died 1771. Three children, Mary, Isabella and Phebe. See Chalkley, Vol. III, p. 122.

II. JAMES was living in Rockbridge 1772. Is said to have emigrated to Georgia.

III. NATHANIEL, born about 1747. Was living in Rockbridge 1768.

IV. DOROTHY, probably married David Dryden. Sons, Nathaniel, William and Thomas.

V. MARY, married Joseph Reed, of Rockbridge County.

VI. HANNAH, married John Smiley, of Rockbridge Co.

VII. ALEXANDER, born 1749, died about 1765.

VIII. THOMAS, born 1753. Botetourt County. Was in battle of Point Pleasant 1775 in Capt. John Murry's company. Probably emigrated to Mecklenburg County, N. C. See McClures in N. C.

IX. MARGARET, born 1757, married a Lee. Emigrated to the South.

X. MOSES, born 1760. Probably the Moses McClure 1790 in Mecklenburg County, N. C. See McClures in N. C.

E. SAMUEL MCCLURE, wife MARY, possibly a Kelso. He was probably the oldest son of Halbert and Agnes Mc Clure, and the first to settle in the county, as Halbert's farm joined his. He was a member of Capt. John Buchanan's company of militia 1742. He died 1779, and his will is recorded in Lexington, Va. Nine children:

I. SAMUEL MCCLURE lived in Forks of James River.

II. WILLIAM MCCLURE, m. Dec. 26, 1769, Jean Trimble, dau. of James Trimble and Sarah Kersey of The Cowpasture. His deed for 274 acres in Forks of the James, cor. to Samuel McClure, is dated 1769. He signed a call for Rev. Wm. Graham to Monmouth church. He died 1785. His will is recorded at Lexington, Va. Eight children:

 1. WILLIAM, m. Jan. 20, 1790, Mary Shields, dau. of Jane Shields, a widow. Em. to Woodford Co. Ky.

 2. JAMES, d. s. 1827. Will, Lexington, Ky.

 3. SAMUEL, soldier of War of 1812.

 4. ALEXANDER, soldier of War of 1812. Wife, Betty was living 1827.

 5. JOHN, m. 1st Jane ——; 2nd Nancy ——; died in Rockbridge County 1834. Four children:

  (1). James Madison, b. about 1810, student at Washington College 1828. D. S.

  (2). John Trimble.

  (3). William Franklin.

(4). Eglantine, m. Addison J. Henderson.
6. SARAH. No record.
7. MARY. No record.
8. AGNES. No record.
III. ALEXANDER McCLURE. No record.
IV. ANNE McCLURE. No record.
V. AGNES McCLURE m. James Campbell.
VI. ELIZABETH McCLURE. No record.
VII. HANNAH McCLURE. No record.
VIII. JEAN McCLURE m. an Elliot.
IX. MARY McCLURE m. a Ratliff.

Alexander McClure, of Eastern Virginia, doubtless belongs to this family. He m. Nancy Dupuy. Seven children: Abram, Mary who m. a Campbell, Samuel, Alexander who m. Webb, William, Bartlett who m. an Ashby, and Achsa who m. a Bacy. (See Virginia Historical Soc. V. 177).

Cora T. McClure, who m. Henry Craig Ewell May, 1873, and died March 25, 1874, probably also belonged to this line.

LINE OF JOHN McCLURE.

JOHN McCLURE, the brother of Halbert, died prior to 1748, intestate. He was doubtless the John McClewer who, with two Alexander McClewers, Halbert and Moses McClewer, were members of Capt. McDowell's company of militia, 1742. His son,

A. Halbert McClure, nephew of Halbert, Sr., was born about 1738. Possibly the Halbert McClure who settled on the Holstein, 1793, and ancestor of Rev. Arthur McClure, a Methodist minister, born in East Tennessee Feb. 16, 1801, entered the Tennessee Conference 1822, and died September 26, 1825. "A young man of much promise, excellent in abilities and graces, and an eloquent and successful minister."—Conference Minutes, vol. I, p. 550.

It is my opinion that the following McClures, who settled on lands adjoining the sons of Halbert McClure, were also sons of John.

B. ARTHUR McCLURE. His deed from Benj. Borden, 1749, for land on Mill Creek corner to David Dryden.

He married, about 1750, Frances ——, possibly a daughter of John and Mary McNabb. A son, Arthur McClure, born about 1752, added to list of tithables, 1768. The father of John Arthur McCluer, born in Rockbridge County. He married Isabella McCorkle of Rockbridge and settled about 1775 on James River, near Buchanan. He died 1854. Six sons and three daughters, among them Capt. John A. McCluer, who married his cousin, a Miss Wilson, of the Rockbridge family; parents of Mrs. N. J. Baker, of Nace, Botetourt County, Va.

C. JAMES McCLURE. A deed for 200 acres "on a branch of James River called the Mary," 1748. He was living, 1790, in Amherst County. A son, John McClure, grandfather of Esther McClure, who married Elijah Goodwin, grandparents of Mrs. Stanley Beasley, of Petersburg, Va.; of Lucy McClure, who married a Pugh, of Charlottesville, Va.; and of John McClure of Nelson County, who married a Slaughter and emigrated to Bloomington, Ill.

D. JOHN McCLURE. His farm on James River cornered Moses McClure and Thomas Paxton. He was a member of Timber Ridge, signing the call 1753 for Rev. John Brown. He died about 1780. His wife Catharine signed the call 1789 from Monmouth church for Rev. Wm. Graham. A son,

I. JOHN McCLUER, born about 1750 and died 1822. His will is recorded at Lexington, Va. He lived four miles from Lexington and four miles from the junction of the North and James Rivers. He m. about 1775 Nancy Steele. Seven children:

1. ARTHUR McCLUER, farmer, lived at Fancy Hill, Rockbridge County; died 1855. He m. Nancy Edmondson. Five children:

(1). Dr. John Edmondson McCluer, b. 1798, united with Monmouth Presbyterian church Sept. 23,

1822, and d. at Richmond, Va., Nov. 13, 1873. Alumnus Washington College; surgeon C. S. A. He m. Martha Parry, of Rockbridge. Son:

a. Charles E. McCluer, b. 1836 and died at the home of his son, Chas. P. McCluer, Tarboro, N. C., June 14, 1914. Electrician. For many years an Elder in the Third Presbyterian church, Richmond; Va.; lived later in Norfolk, Virginia.

(2). Paxton McCluer, d. s.
(3). Dr. David McCluer, d. s.
(4). Sally, m. 1st, 1825, William McClure, Jr., Woodford County, Ky. See Rockbridge County records. M. 2nd a Craig.
(5). Robert Campbell McCluer, b. 1816, d. in Rockbridge County in 1881. Farmer, Ruling Elder in the Falling Springs Presbyterian church, where he is buried. He m. Mary Parry. Eight children:

a. Arthur D., Alumnus Washington College; Confederate soldier 27th Virginia Regiment; killed at Malvern Hill July, 1862.

b. John Parry, Alumnus Washington College, Superintendent of Schools, Buena Vista, Va. An Elder in the Buena Vista church. He married Emma Steele. One daughter:

(a). Isabelle McCluer, who m. John Alexander Stuart, an Elder in the Buena Vista church. They have two children:

  a. John Alexander, Jr.
  b. Parry McCluer.

c. Louisa V., m. Samuel Gilmore. Six children, viz: William, Charles, Evelyn, Edward, John and Nancy.

d. Nancy E., m. David E. Laird. Five children: David, Edwin, Frank, Parry and Robert.

e. Martha J., m. James Harry Gilmore. Four children: Arthur, Rev. Robt. C. Gilmore, pastor

Presbyterian church, Fredericksburg, Va., Thomas and William.

f. Sarah H., m. Rev. H. R. Laird. Five children: Mary, Harvey, Henry, Lilla and Arthur.

g. Roberta E., m. Joseph S. Paxton. Two children: William and Robert Paxton.

h. Lilla K., m. Rev. Alex. F. Laird. Son, John Laird.

2. JOHN STEELE McCLUER, Captain of a militia company 1812. Farmer. Lived at Locust Grove, the homestead. He m. first a Haven, born and raised near Christiansburg, Montgomery County, Va.

He m. second, Seges Price Cameron, of Rockbridge Co. Son:

(1). John Grigsby McCluer, Alumnus Washington College. Confederate soldier Twelfth Virginia Cavalry. Lawyer, Parkersburg, W. Va. Two sons:

a. James Steele McCluer, graduate Washington and Lee University. Lawyer Parkersburg, W. Va.

b. John Cameron McCluer, graduate Washington and Lee University. Lawyer Parkersbury, W. Virginia.

3. NATHAN McCLUER, b. Sept. 11, 1789, and died Aug. 8, 1855. Farmer, living on Buffalo Creek, five miles from Lexington. An Elder in Falling Springs Presbyterian church. He m. Feb. 22, 1821, Jane McChesney (1793-1845). Two children:

(1). Nancy Jane, Nov. 8, 1821–Aug. 8, 1855; m. Jonathan Poage Lackey.

(2). John William, b. March 21, 1830, d. Aug. 7, 1882. He m. September 4, 1855, Elizabeth Catharine Shafer. Five children:

a. Robert Shafer, Alumnus W. and L. University. Living, Roanoke, Va.

b. Charles Christian, Alumnus W. and L. University. Farmer, Cherokee, Tex.

c. William Bittinger, W. and L. U. Chicago, Ill.

d. Hugh Brock, W. and L. U. Farmer; m. Eva Steele. Dead.

e. Frank.

4. ROBERT MCCLUER. Physician, surgeon U. S. Army. Alumnus Washington College. United with Monmouth church Oct. 1, 1820. Emigrated to St. Charles County, Mo., 1829, where he died Sept. 21, 1834.

He m. Sophia Campbell, dau. of Prof. Samuel Campbell, of Washington College, and Sally Alexander.

Four children:
- (1). Janetta Campbell, who m. Dr. John Muschany, St. Charles City, Mo. Several children.
- (2). Samuel Campbell, m. Lucretia Fawcett. The following appeared at the time of her death:

MRS. LUCRETIA MCCLUER.

Was born in Harrisonburg, Va., June 2nd, 1822, and died at her home in St. Charles County, Mo., March 10th, 1913, in the ninety-first year of her age.

She was a daughter of Mr. Jos. Fawcett. Her mother's maiden name was Keyes. The Fawcett family are believed to have been of the Huguenot stock. Mr. Fawcett moved to Missouri about the year 1834, and settled at "Old Franklin" in Howard county. Two years later the family moved to St. Charles, one of the old French towns of Missouri. At that time French was the principal language spoken on the streets and in the shops. Here the subject of this sketch passed her girlhood. While yet in school she united with the Presbyterian church in St. Charles. She was married in December, 1841, to Mr. Samuel C. McCluer, of Dardenne, St. Charles county, who died in 1888. He was a son of Dr. Robert McCluer, of Lexington, Va., who moved to Missouri in 1829. From the time of her marriage till her death she continued to reside in the same neighborhood, and was all these years a member of the Dardenne Presbyterian church. She was the mother of ten children, eight sons and two daughters, all of whom are living except one daughter, HENRIETTA, who died in her 46th year. Every one of her children is a member of the Presbyterian church. Two of her sons, UNCAS and WILLIAM, are ministers of the gospel, the former in Arkansas, the latter in Kentucky; two, OSCAR and THOMAS, are ruling elders in the church at O'Fallon, Mo., and two, ARTHUR and LOUIS, are deacons in the O'Fallon and the Dardenne churches respectively. One son, CURTIS, resides near the old home and the youngest, ROBERT, is a teacher in South Dakota. The surviving daughter, SUSAN, married Rev. Wm. McCarty, who died in 1901.

Such a record speaks for itself. Mrs. McCluer was a woman of rare attainments. Her home was ordered with wisdom and fidelity. She was a diligent student of the Scriptures and held with intelligent conviction the system of doctrine taught in the Westminiser Confession of Faith. In the rearing of her family she was careful to inculcate the truths of the Scriptures—endeavoring to instill those sound principles of religion which regulate the life. Her aim was to train her children by precept and example rather than by simple compulsion. She was a life-long reader of good literature, both religious and secular, and was well informed about the great issues in the church and in the State. Her greatest interest in things beyond her own community was in Missions, Home and Foreign. For the great work of propagating the gospel in the world at large she prayed, and to this cause gave liberally of her means. To the last of her extreme age she retained full possession of her mental faculties and talked with interest about the subjects which had so long been foremost in her mind. He death was peaceful—a fitting close to a life of Christian service.

In this age of progressive ideas let us remember that some ideas have been fixed for us in the Word of God, and that among them is that of a true, godly woman. The memory of such a one let us cherish and let us hope and pray that God will continue to raise up those who in the home and in the church will promote His glory.

a. Rev. Uncas McCluer d. at Little Rock, Ark., Aug.

16, 1913. He m. first Charlottee Wakins, of Virginia; second, Elizabeth J. Morgan, of Va. Seven children. "Mr. McCluer's ministry was a long and useful one, and the many whose lives have been ennobled and uplifted by his labors will deeply regret his loss."
(3). Nancy Calhoun, m. Rev. Thomas Watson. Son, Rev. Samuel McCluer Watson, of Howell, Mo.
(4). Robert Alexander McCluer, m. Sophia Ellen Brown. Nine children:

a. Rev. Edwin Brown McCluer, D. D., b. St. Charles, Mo., Dec. 20, 1854. Moderator Synod of Virginia 1906; Associate Editor The Presbyterian of the South. Pastor Presbyterian church, Bon Air, Va. Children: Edwin Alexander, Alumnus W. and L. U.; and Margaret.
b. Clarence Eugene.
c. Charles.
d. Claiborne Davis.
e. Mattie Janetta.
f. Samuel Bascom.
g. Robert Watson.
h. Nannie Sophia.
i. Horace.

Of the three daughters of John McCluer and Nancy Steele,

5. CATHERINE m. Samuel McCorkle, of Rockbridge Co.

6. NANCY, joined Monmouth church August, 1820; m. Dr. Jas. H. Alexander, of Rockbridge.

7. JANE, probably m. a Byars.

LINE OF DANIEL McCLURE.

DANIEL McCLURE was born Inverness, Scotland, May 15, 1765, came to America about 1804 and settled at Harper's Ferry, Va., where he died Feb. 10, 1815. His wife, Elizabeth, died at Harper's Ferry Sept. 27, 1808. Two children:

1. HENRY McCLURE, who settled in North Carolina.
2. DANIEL McCLURE, JR., b. Inverness, Scotland, 1800; m. Jane McKee 1826. He lived in Upperville, Fauquier County, Va., later moved to Rockbridge Co., finally settled in Bath County, where he died, 1884. Jane, his wife, died 1888. Their children were two sons and five daughters.

(1). Asbury C. McClure, killed at Gettysburg 1863. His name appears in the Rockbridge records, Sept. 2, 1861.

(2). William C. McClure, b. Upperville, Va., Sept. 5, 1828. Married 1st, 1851, Margretta McLaughlin, of Rockbridge Co. He m. 2nd, April, 1855, Mary Martha Alexander, dau. of Robt. Alexander, of the well known Rockbridge family. She was born near Wesley Chapel Dec. 14, 1833; d. June 16, 1893; buried at Falling Springs church. Wm. C. McClure d. at Glasgow, Va., June 4, 1907, and is buried at Falling Springs. He was a veteran of both the Mexican and Civil War. Nine children, viz: R. D. and A. A. of Lynchburg, Va.; William A. McClure, Roanoke, Va.; V. B. McClure, San Francisco, Cal.; H. A. McClure, Dallas, Texas; Joseph Scott, E. W., Emma and Nettie, of Glasgow, Va. Joseph Scott McClure is an efficient officer in the Glasgow Presbyterian church.

McCLURES IN BOTETOURT COUNTY.

Territorial overlapping of Augusta, Botetourt and Rockbridge causes confusion in seeking to locate the early settlers of the two latter counties. While there were a number of McClures in Botetourt between 1770 and 1775, most of them lived in that part of the county that was cut off in 1777 in forming Rockbridge and properly belong to the family in that county.

There were one or two families that settled in the present limits of Botetourt.

JOHN and MARY McCLURE settled in the Long Bottom, south side of James River 1764. His wife was an Allen, probably a daughter of Capt. James Allen, of Augusta County, and a sister of Hugh and Malcolm Allen, of Botetourt.

In his will recorded at Fincastle, the county seat, he mentions three sons:

I. MALCOLN McCLURE, m. Elizabeth Evans. He died May 2, 1791. His will is recorded at Fincastle. Two children:

    1. John McClure, who was living in Botetourt 1813.
    2. Mary McClure, who m. prior to 1813, Walker Stuart of Rockbridge, grandfather of Mr. W. C. Stuart, of Lexington, Va.

The widow, Elizabeth (Evans) McClure m. about 1795 Alexander Crawford, of Rockbridge County, an Elder in New Providence church.

II. SAMUEL McCLURE received land from his father 1769. Was about to move out of the State Sept. 23, 1803. Three sons; John the oldest. See Chalkley, Vol. II, p. 89 and 250.

III. NATHANIEL McCLURE, b. 1774, and m. Mary Jane Porter, b. 1773. Emigrated to Grant County, Ky. A son, John Allen McClure, b. in Kentucy 1797, m. Eunice Keeler Fish. Daughter, Laura McClure, b. in Grant Co., Ky., married a Rankin. Judge J. T. Simon, of Cynthiana,

Ky., who m. a McClure, is an authority on this branch of the family.

The records mention MATHEW MCCLURE as one of the appraisers of the estate of Robert Huston September, 1761. (Chalkley III, p. 66). Probably Capt. Mathew McClure, a signer of the Mecklenburg Declaration. See McClures in North Carolina.

WILLIAM MCCLURE, mentioned 1771. Probably the William McClure, brother of Capt. Mathew McClure, of Mecklenburg County, N. C.

The McClures of Lawrence County, Ky., are also from Botetourt county. The following outline was furnished me by Prof. George M. McClure, M. A., of Danville, Ky:

"The branch of the clan from which I am descended, came from Scotland before the Revolution, settling first on the Eastern Shore of Maryland. Later, members moved to Virginia and settled in Botetourt county. My great grandfather, Richard Renshaw McClure, was a soldier in Washington's army. He came to Kentucky after the Revolution and settled in Lawrence county. Two of his sons, William (my grandfather) and Mordecai, served through the war of 1812. One son, John, remained in Virginia and I have no information as to his history. The family Bible of Richard McClure and Mary Crawford, his wife, was for some years in my father's family and the data as to births, marriages, &c., was complete. It was loaned to another member and in some manner lost."

George M. McClure, Editor The Kentucky Standard and professor in the State institution, Danville, Ky., m. a Miss Jasper, of Jessamine County. Four children:

Francis Jasper McClure m. Louisa Batterton, of Danville, Ky., Feb. 18, 1914.

William McClure, a graduate of the State University.

A daughter, graduate of Caldwell College.

A son, not grown.

Dr. William B. McClure, of Lexington, Ky., and Mr. R. C. McClure, a prominent attorney of Louisa Ky., are his brothers.

Possibly CAPT. FRANCIS McCLURE, "who formerly lived at Wheeling Creek," killed in battle near the present site of Pittsburg, June, 1774, and GEORGE McCLURE, Virginia soldiers of the Revolution, belong to this family. See Peyton's History of Augusta County and Dunmore's Wars p. 37. Gen. Geo. M. McClure lived in Bath, N. Y., 1819.

Admiral Schley has written of his McClure ancestors of the Eastern Shore of Maryland.—*Cosmopolitan, Dec., 1911.*

McCLURES IN RICHMOND.

Several of the name appear in the directory. With one exception they are probably descendants of William McClure, of Augusta County.

Mr. E. Mortimer McClure, of the firm McClure, Davenport, Taylor Co., real estate brokers, a well known citizen of the city, writes me that his father, Dr. Robert McClure, was born in Kirkcudbrightshire, Scotland, 1832, and his only surviving children are Robert M. McClure, for several years postmaster at Gordonsville, Va., and himself.

Three of his father's family came to America, viz., a brother, David McClure, who settled in Boston, and two sisters, Margaret and Elizabeth, who lived single in Brookline, a suburb of Boston, Mass.

McCLURES IN WARREN COUNTY.

The last of the name to emigrate to Virginia is represented in Mr. George Cochran Auld McClure, who lives near Front Royal, Warren County, and who very kindly gave me the following information:

WILLIAM McCLURE, his remote ancestor, was born in Stranraer, Wigtown, Scotland, 1698. Married a niece of Viscount Kenmure. Forced to leave Scotland for aiding the cause of Prince Charlie, 1745, he settled in Charleston, S. C. Two sons:

   1. WILLIAM McCLURE, b. about 1742. Two children,
      Janet and Sarah, b, about 1795.

2. COCHRAN MCCLURE, b. 1744 and died Jan. 5, 1820.
Cotton planters and shippers they added to the large estate inherited from their father. Being loyalists they removed to London on the outbreak of the Revolutionary War, returning at its close. During the war of 1812 they returned permanently to London where they died. The family vault may be seen in Bunhill Fields.

Cochran McClure left three children, James, John, and William who was born June 29, 1790, and died of yellow fever in the West Indies 16th May, 1832. He m. his cousin, Janet McClure, 23rd May, 1823. Two children:

   (1). JANET MORLAND MCCLURE.
   (2). WILLIAM BAINBRIDGE MCCLURE, b. in London 1824; educated at the Royal Naval School, Eng.; resigned and went to the West Indies. Recovered from yellow fever and settled in Alexandria, Va., where he m. 5th Oct. 1848, Ann Auld. He died at St. Paul, Minn., 15th March, 1904. He was a great traveler, having several times circumnavigated the globe. Three sons:
  a. Charles McClure, of Sidney, Australia.
  b. George Cochran Auld McClure, Arco, Va.
  c. John McClure, married. Children, Janet Morland, John, Merrill and others.

McCLURES IN

NORTH AND SOUTH CAROLINA.

It is morally certain that the families in North Carolina and Virginia were related. The first mention of the name is RICHARD McCLURE in petition Nov. 22, 1744, for 300 acres of land in Currituck County. He was doubtless a brother of James and John McClure, of Chester County, Penn. It is stated in Clarke's Colonial Records that on Dec. 4, 1744, he was paid one hundred pounds for his services as Clerk of the Committee of Public Accounts at Bath.

"April 20, 1745.
Gentlemen of His Majestie's Council.

We have resolved that Richard McClure, Clerke of the Committee of Publick Accounts, be allowed forty pounds for acting as Clerke of the said Com. this session, &c.
WILL'M HERRITAGE, Cl'ke Gen'l As'bly."

It is certain he did not remain in Currituck, but moved either to the western part of the State or back to Penn.

JOHN McCLURE died in Mecklenburg County, 1778. His will is recorded at Charlotte, book B, p. 57. Son, Joseph and a brother, Charles. Jas. Montgomery and Wm. Mc-Lure, witnesses.

The census of 1790 gives the following:

In Burke County, 1790.

A. ANDREW McCLURE, two sons over sixteen, five under sixteen and four daughters.

B. FRANCIS McCLURE, two sons over sixteen, three daughters. He is doubtless the Francis McClure, Revolutionary soldier, who enlisted in 1777.

In Rutherford, an adjoining county, 1790:

A. JOHN McCLURE, two sons over sixteen, two daughters.

B. JOHN McCLURE, three sons, all under sixteen.

C. RICHARD MCCLURE, two sons over sixteen, three under sixteen; two daughters. Revolutionary soldier, pensioned 1830.

In Mecklenburg County, 1790:

A. JOHN McCLURE, JR., three sons over sixteen, one under sixteen; two daughters.

B. JOHN McCLURE, JR., two sons under sixteen, three daughters. Probably son of Mathew.

C. MOSES McCLURE, son of Thomas; two sons under sixteen, two daughters.

D. MOSES McCLURE, JR., single. Probably of the Rockbridge family. See p. 140.

E. CAPT. MATHEW McCLURE.

F. THOS. McCLURE, SR., one son over sixteen, one under sixteen, one daughter. Revolutionary soldier, ensign, wounded and pensioned. See p. 131.

G. THOMAS McCLURE, JR., married. No children. Probably son of Mathew.

H. "WIDOW" McCLURE, one son over sixteen, one under sixteen; one daughter. Possibly the wife of Captain John McClure, of South Carolina, who died in Charlotte 1780.

I. WILLIAM McCLURE, two sons over sixteen, two under sixteen; five daughters. Probable brother of Captain Mathew McClure.

In Orange County, 1790:

HENRY McCLURE, two polls and 1,230 acres of land.

JOHN McLURE died in Mecklenburg County 1817. His will is recorded, Book E, p. 21, Charlotte, N. C. He married about 1774, ANN McKEAGAN. Six children:

I. HUGH, b. about 1775, and died single in Mecklenburg County 1840. His will is recorded, Book H, p. 74, Charlotte.

II. THOMAS, b. 1779, and d. 1860. He m. 1825 Ann Ferris Camfield. Son,
   1. Judge John Joseph McLure, a prominent citizen and Elder in the Purity Presbyterian church,

Chester, S. C. He m. Bettie McIntosh. Several children:
a. J. C. McLure, Chester, S. C.
b. Elizabeth, b. in Chester, m. Paul Hemphill, of South Carolina.

III. WILLIAM.
IV. JOHN.
V. PATSY, died single after 1840.
VI. ANN, died single after 1840.

The best known of the early North Carolina McClures is CAPT. MATTHEW MCCLURE referred to in Wheeler's History of N. C., p. 70, as one of the signers of the Mecklenburg Declaration of Independence, May 20, 1775. In The Mecklenburg Declaration of Independence and the Lives of its Signers, by Geo. W. Graham, M. D., p. 123-4, we read, "In the North of Ireland, about 1725, was born Matthew McClure, where he married; came to America and settled in Mecklenburg County, five miles south of Davidson College about 1751. It is an evidence of his worth that he was chosen one of the delegates to the Mecklenburg Convention of May, 1775. It is not known that he filled any other public position. His house was a rendezvous for the patriots of his section. In January, 1782, the County Court ordered that no person in Charlotte or within two miles of the place, should be permitted to sell any spirituous liquors so long as the hospital was continued in that town and employed Matthew McClure to take possession of all such contraband liquors for the use of the hospital, as the commanding officer should direct. Too old himself to enter active service in the field, his sons were much engaged in the army."

His name is mentioned a number of times in Clark's Colonial Records of N. C.

He died 1805 about 80 years of age. In his will, recorded May 4, 1805. (See Book E, p. 4, Mecklenburg Co.). He disposes of his 1,000 acre farm in Mecklenburg Co.; 800 acres to his son, WILLIAM, and 200 acres to his grandson, Matthew, the son of William; $200 to his son,

THOMAS; $1,000 to his daughter, SARAH, the wife of John Henderson and the mother of Jennet Henderson; bequests to his daughter, MARTHA, wife of Hugh Houston; to his daughter, JANE, w. of Wm. Kerns, and to her four children by her first husband, Geo. Houston; to his d. BETSY, wife of Samuel Harris, and their two children, James and Peggy: to his son, JOSEPH, "if he can be found;" to Matthew, son of his brother, William McClure, deceased of S. C., and to Matthew Morrison, his kinsman of S. C.

He mentions other property in lands west of the Alleghanies, and owned in 1790, six slaves.

The witnesses were William Alexander, Jos. McKnitt Alexander, and J. M. L. Alexander. The administrators, Samuel Harris, Wm. Kerns and Jos. McKnitt Alexander.

This is doubtless the Matthew McClure who was in Augusta County 1760, and who is mentioned only once; one of the appraisers of the estate of Robert Houston. Chalkley III, p. 66.

A family certainly connected with that of Mecklenburg County, N. C., settled some time before the Revolution on Pacolet River, Cherokee County, S. C. The records at Greenville, S. C., give deeds and wills of James, James R., John, Richard, Samuel, Thomas W., Mollie, et al. The best known ancestor is MARY (GASTON) McCLURE, known in the history of South Carolina as "The Heroine of the Cherokee." She was a sister of Dr. Gaston, a Revolutionary patriot and is said to have been born 1725 and died 1800. Four sons in the Revolutionary War.

I. CAPT. JOHN McCLURE, wounded at Hanging Rock Aug. 6, 1780, and died in Liberty Hall, Charlotte, N. C., Aug. 18. Gen. Davis spoke of him as one of the bravest men he had ever known. See Vol. XII, p. 129, "The South in the Building of the Nation." It is also probable that this family is descended from JOHN McCLURE, of Burt, near Londonderry.

He left a son, JOHN McCLURE, who m. Mary Porter; parents of Hugh McClure, who m. Margaret Crain; parents of Eliza Jane McClure, who m. Dr. Abram Da Vega in South Carolina.

II. ENSIGN JAMES MCCLURE, also wounded at Hanging Rock Aug. 6, 1780. He, with his brother-in-law, Edward Martin, while melting pewter to make bullets were captured by Huck and condemned to death. For the full account see McCurdy's History of South Carolina, p. 594.

III. HUGH MCCLURE, Revolutionary soldier. McCrady's History of S. C., p. 762, giving the personnel of the Provincial Congress, 1775, says: "It is at least significant that we find among the returned none of the Brattons, McLures, Hills, Gastons and Laceys who so distinguished themselves when the war of the Revolution rolled back to the upper part of the State."

IV. DR. WILLIAM MCCLURE, mentioned in Wheeler's History of N. C., p. 79, as a soldier of the Revolution, appointed April 17, 1776, surgeon Sixth Regiment; transferred June 7, 1776, to the Second Regiment, Col. John Patten, Commanding. Was captured at the fall of Fort Moultrie, May 12, 1780, and later exchanged.

From a number of letters from him to Gen. Sumter, published in the Colonial records of N. C., we learn that he had an uncle, a Dr. Gaston, killed by the enemy; that all his property in S. C., "which was considerable," had been lost by the war; that his aged mother, who was in affluent circumstances in S. C., had been reduced to poverty by the war; that in the year 1776, in S. C., he was surgeon for the Eighth Virginia Regiment in addition to his own; that he was detained in New Bern, N. C., for some time by reason of ill health.

He was in 1784 appointed one of the trustees and directors of the New Bern Academy. He was on Dec. 29, 1785, appointed one of the commissioners on pension claims.

On Nov. 22, 1785, the Legislature appointed a committee "to examine the model of a boat invented by Dr. McClure, which is represented to be calculated to improve the inland navigation of this State."

In the Senate Journal for December, 1786, "we nominate

Dr. William McClure, &c., &c., Councillors of State," to which office he was elected.

In 1790 the Senate endorsed memoranda submitted by him. He died in New Bern, N. C., 1804.

The New Bern, N. C., records show that he owned a great deal of property in and around the town.

Heitman gives his death at 1825. This is positively wrong, as his will is recorded in Book B, Folio 207, New Bern, written 1794 and proven 1804. Wife, Elizabeth. Judging from the will he had no sons. He speaks of his brothers and sisters, but not by name. He mentions two daughters, Fanny Bachelor and Hannah, who was not to marry until she was twenty and to live with Margaret Gaston. See Heitman, p. 275.

The name is found in various places in North and South Carolina. Paul Wheeler McLure, Spartanburg, belongs to one of the original families. Also Rev. Daniel Milton McLure, b. Flat Rock, N. C., 1835; graduate Davidson College and Oglethrope University 1858; Columbia Theological Seminary. Ordained 1864 and died 1865.

Thomas Henry McClure, Jr., of Charleston, S. C., is a descendant of DAVID McCLURE, who came from Londonderry, Ireland. Had a son, William John McClure, father of James and William McClure now (1913), living in Chattanooga, Tenn., and Thomas Henry McClure, Sr., his father. Misses Emily and Margaret McClure, of Charleston, also belong to this family.

Doubtless the best known to-day of the family in the Carolinas is REV. ALEXADER DOAK McCLURE, D. D., the beloved pastor of St. Andrew's Presbyterian church, Wilmington, N. C., since 1891.

His grandfather, WILLIAM McCLURE, was born near Ballemony, County Antrim, Ireland; was brought when three years old to America by his parents, who settled in Tennessee about 1790. He married at Greenville, Tenn., later moving to Marshall County.

Rev. Alexander Doak McClure was b. at Lewisburg, Tenn., July 9, 1850, and was named for his father's friend,

Rev. Alexander Doak, of p. 126. Graduated at Princeton University 1874, and Princeton Seminary 1879. Ordained 1878. He m. 1888 Roberta Calloway, of Louisville, Ky. Two children: Edwin McClure, graduate Davidson College, and Elizabeth McClure.

A brother, Robert G. McClure, is married and lives in Indianapolis, Ind. A sister, Mrs. John B. Knox, in Anniston, Ala.

This family is probably related to that of S. S. McClure of McClure's Magazine.

Rev. H. E. McClure, a retired Presbyterian minister living in Waynesboro, Ga., states that his grandfather, James McClure, came to Georgia from South Carolina many years ago. The family, a large one, settled first in Maryland; some of them em. West. His family was related to that of the late Hon. A. K. McClure, LL. D., of Philadelphia.

McCLURES IN MISSISSIPPI.

The only family in the State known to the writer, is that of JAMES McCLURE, a prominent citizen and merchant of Fayette, Jefferson County. Alumnus Washington and Lee University, 1877. Delta Psi fraternity. Assistant Professor Jefferson College, Miss. His father was born in Campsie, Scotland, where his grandfather was a school teacher.

McCLURES IN MASSACHUSETTS.

Two sons of John McClure, of near Londonderry, Ireland, probably the John McClure, Ruling Elder in the Congregation of Burt, 1700, came to Boston in 1729.

A. DAVID MCCLURE removed to Brookfield, where he left a numerous posterity. Among them—

I. DAVID MCCLURE, JR., (1735-1813). Surgeon in the Revolutionary War. He m. Lucy Kibbe. Two sons:
    1. DR. DAVID MCCLURE, JR., physician, Stafford, New York.
    2. SAMUEL MCCLURE, m. Nancy Caroline Calhoun. Three children:
        (1). Augusta, m. George W. Archer.
        (2). Mary, m. Henry C. White.
        (3). Wm. H., m. Olive Merrill, dau. Olive C. McClure, b. Cedar Falls, Iowa, and m. Frederick Markley.

B. SAMUEL MCCLURE, the other son, remained in Boston, and was the first Elder of the First Presbyterian church of Scotland in Boston, elected July 14, 1730. His children were: JANE, m. Rob't Fullerton; DAVID, drowned at sea; ANNA, m. Matthew Stewart; SAMUEL, m. Martha McClure; MARGARET, m. Thomas Stinson; JOHN, m. Aug. 5, 1740, Rachel, d. of William McClintock, of Londonderry, who came with the McClures to America. McClintock, with his parents, suffered the horrors of the siege of Londonderry.

John McClure d. Aug. 30, 1769.

Five children:

1. WILLIAM, b. Sept. 3, 1741, m. first, Martha Weir; second, Tammy Burns, both of Boston. Was Lieut. of a Privateer, Boston, Revolutionary War. Captured and died at sea 1783, returning from England.

2. SAMUEL, b. July 6, 1743. Captain, Revolutionary War; commanded a company of militia at Ticonderoga 1777. Died July, 1815, at Concord, N. H. Married, first,

REV. DAVID McCLURE, D. D.,
1748-1820.

Abigail Dean, of Exeter; second, Miriam Dalton, of Haverhill.

3. JOHN, b. March 3, 1745. Major, Revolutionary War, Ga. regiment, died, Boston, May 18, 1785. He m. first, a Davis, of Savannah, Ga. Second, SARAH DAVIS, dau. of Jas. Davis, of New Bern, N. C. He is mentioned in the New Bern records 1782. One son,

  (1). James Davis McClure, mar. a Smith, of Cape Cod, Mass. He died at sea, 1808. Two sons,
  a. Jesse McClure.
  b. James Henry McClure, b. in the first house built in Washington, N. C., June 4, 1808. He married Louise Ellis, of near Greenville, Pitt County, N. C. He died Nov., 1902. A highly respected citizen. Ten children, viz: Mary Lurana, Emma Felicia, Oliver Hunter, Susan Matilda, Louisa, George Ellis and John Frederick Latham. Three others, a daughter and two sons, died infants.

John Frederick Latham McClure was born July 8, 1855, m, Anna Katherine Habourn, of Washington, N. C., Jan. 29, 1890. They live in Washington, N. C.; members of the Presbyterian church. One child living, James Henry McClure. A second son, Charles Tilghman, d. i.

4. RACHEL, b. Dec. 10, 1746, m. Capt. H. Hunter, a merchant in Boston. Died December, 1813.

5. REV. DAVID McCLURE, D. D., b. at Newport, R. I., Nov. 18, 1748, graduated at Yale College in 1769. Before graduating he formed a purpose to become a missionary to the Indians. The following letter from his mother and father bears on this early purpose:

"BOSTON, July 30, 1764.

DEAR AND LOVING SON—You have greatly rejoiced all our hearts in expressing your zeal and resolution for the glory of God in the service of His Son, Jesus Christ, to carry His gospel among the aboriginal natives. It is the most honorable employment in the world. O, my son, I have given you up to God, soul and body. Many prayers

I have put up to heaven for you. I hope God is answering them now. O, my son, go on in the strength of the Lord and in the power of His might. You may expect onsets from Satan, the World and the Flesh, but the more you find yourself assaulted by them be still more earnest at the Thorne of Grace. The Lord's promise stands sure, "They that seek Me early shall find Me." Give not way to discouragements. Your loving father and mother,

JOHN AND RACHEL MCCLURE."

After four years spent among the Indians, he was compelled to abandon the mission on account of the Revolutionary War. He returned to New Hampshire, was installed pastor at New Hampton church 1776, and at East Windsor, Conn., 1786, where he died June 25, 1820. He wrote several books, among them a history of East Windsor. He is described as "a small man, well formed and with very attractive manners—a man of culture and scholarship." He was a trustee of Dartmouth College, from which he received his D. D. in 1800.

He m. first, 1780, Hannah, youngest d. of Rev. Benj. Pomeroy, D. D., of Hebron County, and secondly, Mrs. Elizabeth Martin, of Providence, R. I. His children were:

 (1). Abigail Wheelock, bap. Sept. 10, 1786; m. Dec. 22, 1801, Oliver Tudor, of East Windsor. Five children.

 (2). Rachel McClintock, b. Oct. 29, 1783; m. Nov. 27, 1806, Elihu Wolcott, of East Windsor.

 (3). Mary Ann; died July 12, 1789.

 (4). Susannah Willys, bap. Nov. 16, 1788; d. s. aged about 35.

 (5). Hannah Pomeroy, bap. Aug. 28, 1791, died Aug. 25, 1804.

The Diary of Dr. David McClure, an exceedingly interesting book and containing an outline history of his family, was published 1899 by John P. and William B. Peters, of New York.

 6. JAMES, b. Feb. 25, 1750. Mar. Eliz. Randlet, of

Exeter, N. H. Capt. and owner of a merchant ship. Died in Dublin, Ireland, March, 1791.

7. DANIEL, b. March 13, 1753. Died in Savannah, Ga., Sept. 15. 1775.

8. THOMAS, b. Nov. 21, 1754; m. first, Nancy Hunter, Bristol, Me. Second, Mary Wilson, of Boston. He, like his father and grandfather, was an Elder in the Federal Street, the first Presbyterian church of Boston. He left the church when it became Unitarian under Dr. Channing. Two greatgrand-sons now live in New York City, viz: Wm. R. Peters, a distinguished lawyer, and Rev. John P. Peters, rector St. Michael Protestant Episcopal church.

9. JANE, b. July 27, 1757. Married James Randlet, of Exeter. Died about 1805.

10. NANCY, b. Aug. 5, 1759. Died in Boston.

11. JOSEPH, b. Sept. 3, 1761. Farmer, married and lived in Bedford, Maine.

12. BENJAMIN, twin, b. Sept. 3, 1761. Sea captain. Died at Exeter Feb. 18, 1787.

13. RUTH, Dec. 26, 1763—Oct., 1765.

Doubtless belonging to this family was Rev. Alexander Wilson McClure, D. D., b. Boston May 8, 1808; educated at Yale and Amherst Colleges and Andover Theological Seminary, class 1830. Pastor and Editor, died Sept. 20, 1855. "Dr. McClure was truly a learned scholar, a genuine wit, keen dialectician and a practical controversialist. Ardent and honest as the sunlight, abounding in good feeling and simple in manners as a child, he was a man of positive convictions, fearless of consequences in the advocacy of truth and in assailing popular error. Yet with all his exuberant mirth and knowledge of the world, Dr. McClure was pre-eminently a devout and humble Christian minister."—Sprague's Annals of the American Pulpit, Vol. II, p. 7.

Another living member of this family, is Arthur G. McClure, of New York City.

McCLURES IN NEW HAMPSHIRE.

Prof. Charles F. W. McClure, of Princeton University, kindly gave me the following outline of his family:

A. DAVID McCLURE, founder, was born about 1720; emigrated from the north of Ireland, possibly Londonderry, to Boston, 1740, thence to Chester, N. H., and in 1743 located in Candia, N. H., then, or a little later, called Charmingfare. His home was a mile or two from the Green in Raymond. He married in Ireland, Martha Glenn, who came with him to Boston. He perished in a snow storm about 1770.

They left three children:

I. ELIZABETH, born Nov. 25, 1738, m. Oct., 1761, Alexander McClure, b. 1734.

Twelve children:

1. JANE, or Jean, b. 1763, d. i.
2. JANE, b. June 2, 1765, m. Ezekiel Fullerton of Raymond, N. H. Issue—
   (1). John M., (2) James, (3) Betty. About 1793 the family moved from Raymond to Cambridge, Vt.
2. MARTHA, b. June 31, 1766, m. —— Smith.
4. MARY, b. Dec. 1768.
5. JAMES, b. June 9, 1771, d. s.
6. ALEXANDER, JR., b. Oct. 11, 1773, d. Feb. 8, 1850.
7. BETTY, b. June, 1775, d. i.
8 and 9. Twins, b. Oct. 1777, died at birth.
10. ELIZABETH, b June 19, 1780, m. Jonathan Nay, of Georgia, Vt.
11. HANNAH.
12. FANNY, b. April 24, 1784, d. July 5, 1815.

6. ALEXANDER McCLURE, JR., sixth child of Elizabeth and Alexander McClure, Sr., married, first, Sarah Nay, of Raymond, N. H.. Nine children:

(1). Samuel, m. Mary Gilman, of Raymond, N. H.
three children: H. G. McClure, T. F. McClure, a daughter who m. S. B. Gove.
(2). James, d. in N. Y.

(3). Thomas, b. 1830, d. April 28, 1832.
(4). David, settled in Cambridge, Mass., d. Jan. 20, 1852. Married Sarah Burrage.
(5). John Nay.
(6). Sarah, d. Nov. 28, 1912.
(7.) Elizabeth, m., first, Moses Hoyt, of Raymond; second, Rev. B. G. Manson.
(8). Abigail, d. i.
(9). Mary, d. i.

Alexander McClure, Jr., m., second, a cousin, Martha Varnum. They had nine children:
(10). Moses, died in California, 1858.
(11). Alexander, 3rd, died in California, 1858.
(12). Martha Glenn, died in Raymond.
(13). Frederick, died in Raymond.
(14). Jesse.
   Four other children died in infancy.
(5). John Nay McClure, fifth child of Alexander and Sarah (Nay) McClure, m. Mary, daughter of Isaac Brown, of Fremont, N. H. Three children:
a. Charles Franklin McClure, b. in Raymond, N. H. 1828; was living Nov. 16, 1913. He m., 1851, Joan Elizabeth Blake, daughter of Sherburn Blake, of Raymond, N. H. Five children:
(a). Mary Louise, m., first, Edward Custer; m., second, T. S. Ellery Jennison, of Boston.
(b). Elizabeth Pierce, m. E. V. Bird, of Boston.
(c). Arabella Hersey; single.
(d). Charles Freeman William, Professor of Comparative Anatomy, Princeton, N. J.; single.
(e). Ethel Melvina, m. Dr. Edward Briggs, of Boston.
b. John Freeman; died single.
c. Susan Melvina, m. Seth M. Williams.

II. MARY, second child of DAVID McCLURE and MARTHA GLENN, m. Thomas Patten of Candia, N. H. They had 12 children:

1. Elizabeth, mother of Martha Varnum, second wife of Alexander McClure, Jr.
2. Martha; 3. Sarah; 4. Rachel; 5. Margaret; 6. Hannah; 7. Ruth; 8. Mary; 9. Joan; 10. Samuel; 11. Moses (father of Rev. Moses Patten, of Rochester, Vt.); 12. Thomas (father of Thomas Patten, of Raymond).

III. JAMES, third child of David McClure and Martha Glenn; m. Mehitabel Burpee, of Candia, N. H. They had seven children:

1. Elizabeth, b. Feb. 26, 1770; m. Dearbon, of Woodstock, N. H.
2. James, b. Sept. 1, 1771; died infancy.
3. Mehitabel, b. Jan. 31, 1774; m. Bursel, Candia, N.H.
4. James, b. Nov. 4, 1776.
5. Rebecca, b. Feb. 13, 1780; m. Hall, Candia, N. H. mother of Orrin Hall, of Cambridge, Mass.
6. Nathaniel, b. April 9, 1785; went west when young.
7. Sally, b. July 19, 1792; m. Thomas Patten.

Other McClures in New Hampshire were—

DAVID McCLURE, b. Goffstown, N. H., 1758, and died 1835. He was a Revolutionary soldier, Sergeant to Capt. John Duncan. He m. Martha Wilson. Son, Manly W. McClure, who m. Martha M. Page; parents of Mary McClure, wife of James Clark.

CAPT. JAMES McCLURE, b. in Londonderry, 1753. Settled in New Hampshire. Revolutionary soldier. Captain 4th Continental Artillery. Died 1840. Married Mary Nesmith.

McCLURES IN NEW YORK.

REV. JAMES GORE KING McCLURE, D. D., President of the McCormick Theological Seminary, Chicago, was b. at Albany. N. Y., Nov. 24, 1848, and graduated at Yale College 1870, Princeton Seminary, 1873.

Under date of Oct. 22, 1910, he gave me the following statement: "My own branch of the McClures came to Al bany, N. Y., in 1801, from the north of Ireland. We never have been able to connect ourselves with any other branches of the McClures in America, nor have we been able to ascertain anything about the original family in the north of Ireland. My grandfather's name was Archibald McClure, and his wife, Elizabeth Craigmiles. There are various family traditions about the character and life of my ancestors in the north of Ireland, but they afford no definite historical basis for the ascertaining of a geneaological line."

I learn from another source that Archibald McClure, before emigrating to America, resided in or near Belfast. There is a family tradition that his ancestors in Scotland, during the persecution of the Covenanters, were once hidden under a load of hay, into which the soldiers thrust their weapons, but without doing injury to those concealed. See page 13.

There is a village near the Delaware River in Broome County, N. Y., called McClure Settlement, probably of the Pennsylvania family. There are a number of McClures in New York City, among them a prominent law firm, David and John McClure.

SAMUEL SIDNEY McCLURE, founder of McClure's Magazine, was born at Drumaglea, County Antrim, Ireland, Feb. 17, 1857. He states in his Autobiogrophy (1913) that his family came to Ireland from Galloway, Scotland, about two hundred years ago. It is stated elsewhere that his remote ancestor was Daniel McLewer, supposedly descended from the Huguenot family of De la Charois, of noble French extraction which claims descent from John, Duke of Gascony. This Daniel may be the same Daniel McLewer who was an Elder attending Templepatrick Presbytery, Ulster, 1738.

His mother, Elizabeth Gaston, descended from a French Huguenot family that came to Ireland after the revocation of the Edict of Nantes. Mary Gaston, the mother of Dr. William McClure, who died at New Bern, N. C., 1804, doubtless belonged to this same stock.

S. S. McClure, speaking of his grandfather, Samuel McClure, said: "He was a man so constituted that he not only would not yield in opinion—he could not. I believe before changing his mind on a point on which he had determined, he might have been tied to the ground and cut to pieces inch by inch."

Rev. William Ramsey, in his letter published in the Autobiography, says of WILLIAM McCLURE, brother of Samuel, "There are no doubt many of his equals in honesty and principal, but none could exceed him or his family, or indeed any of the McClures. He is often in my mind, not only as a devoted Christian, but as so upright in word and deed that, when I lost him I knew of none to fill his place in my heart."

Samuel McClure had seven sons, among them Thomas McClure (1832-1860). His widow, Elizabeth (Gaston) McClure, with her four sons, Samuel, b. 1857, John, b. 1858, Thomas, b. 1860, and Robert B., came to America, settling in Indiana, 1866. Robert B. McClure died at Yonkers, N. Y., May 30, 1914.

Mr. Hugh S. McClure, with the American Exchange National Bank, New York, belongs to a family that lived at Dernock, near Bollymoney County, Antrim. His father, Rev. Samuel McClure, ministered at Cross-roads, near Londonderry, where he died 1874. Has a brother, Rev. John J. McClure, D. D., Capetown, South Africa.

McCLURES IN PENNSYLVANIA.

The first mention of the name in America is in Pennsylvania. The family is probably more numerous there to-day than in any any other State in the Union, while descendants are to be found in every section of the country. The earliest record is that of ROBERT McCLURE, in Dauphin County, 1722.

McCLURES IN CHESTER COUNTY.

Four brothers settled in Currituck County, N. C., about 1740. Not finding the climate healthful, James and John emigrated to Pennsylvania and settled in Uwchlan Township, Chester County. Their deed is dated Oct. 12, 1748. Of these brothers:

A. JAMES McCLURE m. Mary Lewis and left a daughter, ESTHER. No further record.

B. JOHN McCLURE, b. 1705, m. about 1743, Jane Ahll, and died March 25, 1777. JANE, his wife, died Feb. 15, 1762. Four children:

I. ESTHER, b. Sept. 10, 1744, m. —— Williams.

II. CAPT. JAMES McCLURE, a Revolutionary soldier with Gen. Proctor, b. Jan. 11, 1746; m. his first cousin, Esther McClure, daughter of James and Mary Lewis McClure. Four children, viz: Jane, Rachel, Mary and Silas. Silas married and left a son, James, an Elder in the Nantmeal Presbyterian Church, 1870.

III. MARY, b. Oct., 1747, d. s.

IV. JOSEPH, b. Oct. 27, 1749, m. Martha Thompson, of Uwchlan Township; died Oct. 15, 1827. MARTHA, his wife, died Nov. 23, 1829, aged seventy-three years. Eight children, viz: Jane, Elizabeth, James, Joseph, Martha, John, Rachel and Mary. Of these,

3. JAMES, lived near Landisburg, Perry County. One of the first Elders in the Landisburg Presbyterian Church, 1823. Died after 1852. He m. Hannah McKay. Son,

    (1). William McKay McClure, one of the first Elders in the Bloomfield Presbyterian Church, Perry County, 1834. Two sons.

        a. Wiliiam R. McClure, m. Ida Coulon. Died 1899.

        b. Charles V. McClure, dealer in Real Estate, Greene, Iowa, and to whom I am indebted for the information of this branch of the family. He was born August 24, 1845. Soldier in the War between the States; private Co. H, 49th Penn. Vol. Infantry, 6th Corps, Army of the Potomac.

Two sons—

    (a). Charles A. McClure, was private in the 49th Iowa Vol. Infantry, Spanish-American War. Promoted by President McKinley to 2nd Lieutenant in the regular Army. Resigned on close of hostilities in the Philippines.

(b). James Barrett McClure was floor clerk (1912) of the United States Senate.

4. JOSEPH, was ordained an Elder in the Brandywine Manor Church, 1830.

6. JOHN McCLURE, b. in Chester County July 26, 1791; m., first, Feb. 6, 1816, Elizabeth Mackelduff, of Honeybrook. She died August 22, 1822. Two sons.

   (1). Dr. Joseph M. McClure, ordained an Elder in the Nantmeal Presbyterian Church, Chester County, 1870. He died some years ago, leaving a widow and two daughters, viz: Margaret, living at Lyndell, Chester County, and Mrs. Wm. Pemrock, Atlantic City, N. J.

   (2). James McClure.

He married, second, January 13, 1824, Elizabeth Mackelduff, a first cousin of his first wife. Three children.

   (3). Elizabeth, m. Robert Neely.
   (4). John, who married and left a daughter, Mrs. Adda B. McSparran, Peter's Creek, Lancaster County.
   (5). Samuel M., d. s.

He died Feb. 9, 1873. Elizabeth, his wife, died Dec. 15, 1867, aged seventy-three years. The address of Rev. A. Nelson Hollifield on the occasion of his death, delivered at the Fairview Presbyterian Church, Wallace, Chester Co., Feb. 13, 1873, is preserved in book form. The following is from this address:

"The McClure family, ever since its settlement in the American colonies, has been highly respectable. John, (the grandfather of the deceased) and all of his family, were persons of superior intelligence. They were well-to do in wordly possessions, industrious, honest and economical. During the period that preceded and succeeded The Declaration of Independence, they were warm and active partisans of the American cause. From conviction, they were Federalists, and espoused the principles of George

Washington, Alexander Hamilton and John Jay, as opposed to those of Thomas Jefferson, Aaron Burr and James Madison. Strange to say, with but two exception, all of the decendants have adhered to the political faith of their forefarthers, being, to-day, republicans. Two of the family, (at least) rendered efficient service in the revolutionary war, James and Benjamin, the eldest and youngest sons of John. The former was commissioned a Captain of Infantry. He was captured by the British near the close of hostilities, and imprisioned at Long Island, from whence he succeeded in effecting his escape, and returned to his father's home. The war terminating shortly afterwards he did not return to his post in the army. Turning from the scenes of war, we come down to more peaceful times. In the year 1812, a poor woman died, leaving two children only a few days old. A neighboring farmer, of means and respectability attended the funeral. He there saw the helpless orphan boys. Who was to care for them? Their father was not in circumstances to permit him to employ a nurse. As the neighbor observed these things, his heart was touched with sympathy. But he did not stop with that. When the funeral was over, he returned to the tenant house where the children were, and having obtained the grateful consent of their father, took them up, one on each arm, and carried them one mile to his home, and presented them to his astonished wife to care for. That man was Joseph McClure, the father of the deceased. When they became older, he sent one of them to reside with the deceased. These boys are now old men. One resides in this township, an honest, sober and industrious citizen. The other lives in the west, a Methodist Minister. According to the records of the Brandywine Manor Church Session, Joseph McClure was an elder there in 1814, but for how long a time preceding that date we are unable to say, as the minute of 1814 is the oldest we could find. December the 9th, 1825, the following minute appears upon the Sessional Records of Brandywine Manor Church: 'Session regret to learn that Joseph McClure, an aged member of Session, being lately stricken with palsy, we cannot expect from him his usual service.' This family has furnished the church with several elders. As we have seen, Joseph, (the father of the deceased) was an elder in 1814. Two of his sons, Joseph and the deceased, were ordained ruling elders, and installed over the congregation about the year 1830, they, together with a number of persons, founded this, the West Nantmeal Presbyterian Church. The deceased was installed a ruling elder here in 1840. Joseph M. McClure, M. D., (son of the deceased) and Jas. McClure, (grandson of James and Esther McClure) were ordained as ruling elders, and installed over this congregation in 1870. In 1872, Joseph M. McClure, M. D., was elected by the Presbytery of Chester a Commissioner to the General Assembly, and was present in that body at Detroit, Michigan. The deceased was one of the most efficient ruiling elders of the two churches with which he was connected. In his younger days he was a very excellent reader. and it frequently happened that, in absence of the pastor, he was called upon to read a printed sermon, which service he invariably performed with great acceptance to the people. He was always in his accustomed seat in church, until the infirmities of age compelled him to have some consideration for the weather. One of his former pastors, Rev. D. C. Meeker, says on this point, in a recent letter: 'He was exemplary, and often self-denying in his attendance upon the services of the sanctuary.'

Another preacher, the Rev. B. B. Hotchkin, D. D.: 'He was devoutly solicitous for the prosperity of the church; free-hearted in service as a member of its Session and Fiscal Board; cordial towards his associates in office, studious of things that make for peace; ever ready to bear his part in its social devotions; lending to the pastor the support of his influence; and as watchful for it as a father for a child. You knew him only when these qualities began to feel the impairing effect of advancing age; I knew him when they were in their vigor.'

No man ever thought more of his church. He was consecrated to its service in youth, and life's setting sun found the veteran of four-score years at the post of duty and of honor. Not only as a ruling elder did he serve the church of his fathers, but rendered efficient service as a trustee for near a half a century. When the first church was built here, he and his brother Joseph subscribed one-fourth of the whole amount required to complete the structure, and when we began to agitate the subject of building this new church edifice, he was the first one to subscribe. He headed the list with one thousand dollars. Subsequently he largely increased that sum. The deceased was a man likely to be misunderstood by strangers. They might consider him harsh, haughty, overbearing. But such was far from the truth. Whatever his naturally reserved manner might indicate, he had a sympathizing heart, was of a very benevolent disposition, and exceedingly kind and friendly. In the social relations of life, he endeared himself by his constancy and affection. He was given to hospitality. His house has long been the minister's home, and nowhere were they more warmly welcomed or generously treated. The deceased was a great reader, but wasted no time on literature of a light character. His Bible, Burden's Village Sermons; The Grace of Christ, by Dr. Plumer, and Religious Experience, by Dr. A. Alexander, (together with the *Presbyterian* and *Evangelist*) completed his reading library. Although he possessed many other valuable works, these were his favorities. Thus he spent the close of his long life, reading religious books and good papers. He could repeat the Shorter Catechism to his dying day, asking and answering the questions himself. By industry he amassed a large property. But that he was rich in faith and good works is more worthy of record. The aroma of the good name he has left behind him is a more inestimable heritage than the fortunes of the Rothschilds, or the wealth of the Astors; a name honorably associated with the Covenanters of Scotland, the battle for priceless freedom on these western shores, and the establishment of the early Presbyterian Church in America."

V. ELIZABETH, b. Oct. 20, 1751, d. s.

VI. RACHEL, b. March 20, 1754; m. John Neal, of Slate Ridge, Lancaster County. Five children.

VII. JANE, b. January, 1757, m. John Wallace, of Honeybrook, Chester County. Son and five daughters.

VIII. BENJAMIN, b. Sept. 9, 1759, d. 1821. Lieut. to Capt. George Crawford, Col. James Dunlop, Revolutionary War. He m. Agnes Wallace. of Chester County. Eight children, viz: Robert, Jane, Mary, Elizabeth, John, William, Esther and James. Mary m. Rev. Wm. Kennedy. Daughter Mary Jane m. Crawford Hindman.

It will be observed that the names of this family are similar to those of a family in Augusta County, viz. John McClure, of Chester Co. Pa., (1705-1777), eight children: Esther, James, Mary, Joseph, Elizabeth, Rachel, Jane and Benjamin.

John McClure, of Augusta County, Va. (1717-1797), eleven children: Anne, Esther, James, Jane, Elizabeth, Martha, Mary, John, Margaret, Andrew and Eleanor.

McCLURES IN CUMBERLAND COUNTY.

The best authority on this and all the Pennsylvania families is Mr. C. P. McClure, of Bunola, Pa. He has compiled a Family History. It is to be regretted that it has not appeared in print, as it doubtless contains much information of general interest. Associated with him in his undertaking is Mr. Roy Fleming McClure, of Chattanooga, Tenn., and Mr. J. H. McClure, of Elizabeth, Pa.

JOHN McCLURE, b. in Scotland 1696, came to Pennsylvania from the North of Ireland 1715; m., 1730, Janet McKnight, sister of John McKnight, Esq., the well known Justice of Cumberland County, Pa. Settled in Cumberland County about 1732, where he died 1757. Eight children, viz:

JOHN. ANDREW, who is supposed to have married Jean, a first cousin, daughter of Abdiel McClure, b. in Glasgow, 1702; son Abdiel, born in Carlisle 1750, ancestor of Rev.

James T. McClure, pastor of the First U. P. Church, Wheeling, W. Va., for forty-nine years.

CHARLES, RICHARD, MARGARET, JEAN, EUNICE and CATHERINE. The above may be the Margaret McClure of Big Spring Presbyterian Church, Cumberland County, who signed, 1786, the call for Rev. Samuel Wilson.

The Richmond, Va., *Standard*, vol. III, p. 7, gives the following:

MARGARET McCLURE, m., April 4, 1776, John Parker, of Cumberland County, b. 1740.

CHAS. McCLURE, b. Carlisle, Pa., 1739, m., first, Emelia who died Feb. 1, 1793, aged 28 years. Two children:

    1. John McClure, m. Jane Blair and left four children, viz: Catherine, Aurelia, Mary, and John, who d. s.

    2. Mary McClure, who m. Joseph Knox.

He m., second, Mrs. Rebecca Parker, widow of Maj. Alexander Parker. She died at Carlisle April 13, 1826, aged sixty-three years. Four children.

    3. Charlotte.
    4. Charles, married Margaretta Gibson. Son, Major Charles McClure of the U. S. army. Sons, George and William McClure.
    5. Judge Wm. B. McClure, Pittsburg, Pa.
    6. Rebecca, m. Rev. F. T. Brown.

Miss Emma McClure, of Elk Lick, Somerset County, belongs to this line.

John McClure, with Andrew Blair and others, was ordained an Elder in the Second Presbyterian Church, Carlisle, 1833; had been an Elder in the First Church.

Joseph McClure, from Carlisle, 1767. Signed the call to Rev. John Steel.

JAMES McCLURE, who settled, 1780 in Newport, Ky., is supposed to belong to the Cumberland family. Died in Newport, 1830.

He m. JANE MILES, who was drowned at Vevay, Ind., March 8, 1818. Six children:

    1. Sarah, m. David Perry; Died about 1815.

2. John, m. Hester Lloyd of Pennsylvania; died of yellow fever at Baton Rouge, La., 1826. Three children, viz: Eliza, James W. and Julia.
3. James H., b. Nov. 13, 1800, m. Mary Lewis; d. in Texas. No chilnren.
4. Eliza, twin, b. Nov. 13, 1800; d. May 24, 1868.; m. Capt. Samuel Perry. Six children.
5. David, drowned at Vevay, Ind, March 8, 1818.
6. Frances S. O., b. June 28, 1803, d. Feb. 4. 1890; m., Sept. 11, 1821, Capt. Samuel Carter. Their youngest daughter, Jane, b. Dec. 20, 1833; m. James B. Stillwell; lived it Seattle, Wash.

Miss Mary E. Applegate, of Chicago, belongs to this line.

Two nephews of James McClure of Newport, from Pittsburg, but of the Cumberland family, emigrated to Illinois about 1845.

"Col. John B. McClure, of Peoria, with his brothers, Robert and Samuel, were from Shippensburg, Cumberland County. He married a lady from Wisconsin. Daughter Mary living 1868. Robert lived in Olney, Ill. Physician. Samuel d. in Pennsylvania."

Col. John D. McClure, of Peoria, was born in Franklin County, Pa., 1835, settled in Peoria, 1849, and died there March 3, 1911.

JUDGE DAVID MCCLURE, b. in Ireland 1726, lived in Cumberland County, Pa.; died in Sherman's Valley, Pa., 1796. Married Jane McCormick. Son, William McClure, b. about 1760, m. Jane Byers. Son, William (1797-1856), m. Margaret Beaver. Daughter, Emily C. McClure, m. William Warmington.

McClure, a village in Snyder County, probably got its name from the Cumberland family. There are several McClures living now in this section—large and prosperous farmers

Col. Alexander Kelly McClure, LL. D., is perhaps the most distinguished of the name in Pennsylvania. Member Penn. Legislature, candidate for Mayor of Philadelphia, personal friend of President Lincoln. Best known as Editor of the *Philadelphia Times*.

He was born in Sherman's Valley, Perry County, Pa., January 9, 1828, and spent his early years on his father's farm. With his older brother he divided his time week about at a country school. In 1846 he made his first visit to Philadelphia in order to get work as a journeyman tanner. He found no work there and tramped to New York, where his luck was no better. He worked his way west until he found himself in Iowa. but still his ill fortune in the tanning trade stuck to him. He then journeyed back east and that fall, in spite of advice to the contrary, went into the printing business, starting with the *Sentinel*, the Mifflin local paper.

At his suggestion an outline history of his family was prepared by Rev. G. O. Seilhamer, Chambersburg, Pa. Col. McClure's sudden death prevented its being published, which is to be regretted, as it doubtless contains much of interest to the family in general.

McCLURES IN LANCASTER COUNTY.

WILLIAM McCLURE, a Covenanter of Dumfries, Scotland, was with his family driven by persecution from his home and country and settled in Ireland. His youngest son, JAMES, emigrated to America, settled, first, in Lancaster County, Pa.; removed in 1772 to Bloomsburg on the Susquehanna, where he built the well known Fort McClure. Revolutionary soldier.

He had five children: Margaret, Josiah, John, Priscilla and James. Margaret, the oldest daughter, m., Dec. 10, 1783, Maj. Moses Van Campen of New York, and died at Dansville, New York, March, 1845. Col. James McClure, the youngest child, was born in 1774, m. in 1796, and died at the old homestead, October 4, 1850. His children were Margaret, James, Mary, Samuel, Eleanor, Josiah, Charles, Priscilla, Benjamin and Alfred. The Rev. Alfred James

Pollock McClure, a clergyman of the Episcopal Church, now living in Philadelphia, belongs to this family.

He and his daughter, Miss Abby McClure, have put into permanent form the record of their branch of the family.

See Penn. Magazine of Biography, vol. XXXI, pp. 504-506 (1907).

McCLURES IN YORK COUNTY.

Miss Martha McClure, Mt. Pleasant, Iowa, is an authority on this branch of the family.

RICHARD McCLURE, from the north of Ireland, settled about 1725 in Paxtang township, then Lancaster, now York county. Four sons born in Ireland.

A THOMAS, d. in Paxtang, 1765. Wife, MARY, d. in Hanover April, 1775. Six children, viz., John, wife Mary; mar. 1775, lived in Mt. Pleasant township; William; Mary, m. Joseph Sherer; Martha, m. Andrew Wilson; Jean, James Burney; Thomas, m. Mary Harvey.

B. CHARLES, wife ELEANOR. He died prior to 1761. Nine children, viz: Arthur, Rebecca, Jennett, William, John, Martha, Eleanor, Charles, Margaret.

C. JOHN, wife MARY. Died in Hanover 1762. Four children, viz: James, William, Jane, who m. Wm. Waugh, Ann. Of these, James was b. 1733, m. Mary Espy and d. at Hanover Nov. 14, 1805. Nine children, viz: James, d. s. Sept., 1815. Martha, m. a Wilson; three children. William, son James. Frances. Isabel, m. Jos. Cathcart. John. Mary, m. Snodgrass. Andrew, m. and em. to Ohio. Six children, viz: John, Hugh, Scott, Andrew W., Ann and Bell. The fourth son, Dr. Andrew W. McClure em. to Mt. Pleasant, Iowa, 1856. Mar. Emily Conaway Porter, a dau. Martha McClure.

D. RICHARD, eight children, viz: Alexander. William, m. Margaret Wright. Jonathan, m. Sarah Hays. Andrew, w., Margaret. Poan, wife Hannah, d. in Northumberland County, Oct. 8, 1833. Margaret, m. Sept. 7, 1757, John Steel. David, m. Margaret Lecky. Katherine, m. Robt. Fruit.

McCLURES IN PHILADELPHIA.

SAMUEL McCLURE, b. in Belfast, Ireland, 1736, d. in Phila. 1790. Revolutionary Soldier. Mar. Jenet Graham. Son,

I. David McClure; m. Ann Russell. Son,

1. Dr. David McClure, m. Eliza Shute Stewart.

(1). Dau. Elise, b. in Phila., m. Henry Payson Gregory; parents of Elise Gregory, b. Oakland, Cal., m. Lloyd Bowman. A number of the name are now living in and near Phila.

Rev. Robert E. McClure, D. D., of Blairsville, Pa., gave us the following information:

His great grandfather, with three brothers, came to Phila. from the north of Ireland.

One of them left four sons, viz: ANDREW. JOHN, who died in West Middletown, Pa.. and whose children all died single. RICHARD. DR. ROBERT McCLURE, b. in Phila. and died in Washington County, Pa. Left a son, Robert Brown McClure, the builder of the first threshing machine in the United States. He m. Letitia Templeton and left nine children, viz: Aaron T. McClure, living in Washington County, Pa.; Rev. Wm. S. McClure, D. D., pastor Second U. P. church, Xenia, Ohio. Dr. James A. McClure, Columbus, Ohio; Mrs. Emma K. MacDill, Middletown, Ohio; Mrs. Alice E. Snodgrass, Pittsburg, Pa.; Miss Etta M. McClure, teacher, Pittsburg, Pa.; Miss Anna L. McClure, West Middletown, Pa.; Mrs. Jas. E. Ralston, West Middletown, Pa.; Rev. Robert E. McClure, D. D., pastor U. P. church, Blairsville, Pa.

The following appeared in the New York Times, Dec. 21, 1913:

A Bible carried under his left arm saved the life of the Rev. Dr. R. E. McClure, pastor of the United Presbyterian Church here and President of the Indiana County Anti-Saloon League, last night, when an assassin's bullet struck the Bible, perforating it and Dr. McClure's clothing.

On his way home from a sick call Dr. McClure was passing two men in a shadowed spot in Stuart street, when he heard a whistle. At the signal one of the men leveled a revolver at the minister and fired. The bullet went wild. A second bullet passed through the Bible and touched Dr. McClure's skin, but did not break it. Unhurt, the clergyman picked up a brick and threw at the men, who fled. One of the men lost his hat, which the minister turned over to the police.

Dr. McClure has been unrelenting in his prosecution of liquor law violators, and to this is attributed the attempt to murder him. He is a Trustee of Westminster College and one of the best known temperance workers in the State.

ROBERT McCLURE, a Revolutionary soldier, lived in Williamsport, Pa. He m. Mary Hepburn. Two children, viz: 1. Hepburn McClure, m. Martha Biles Anthony, d. Annie Rachel. 2. William McClure, m. Hannah Smith; son Edwin Parson McClure, m. Elvira Grier, dau. Margaret, b. in Bushville, Pa.

JOHN McCLURE, who died in Morgantown, W. Va., 1874, doubtless belongs to the Pennsylvania family. He m. 1835, Martha Steele (1809-1910), b. in Greene County, Pa., dau. of John Steele, b. in Augusta County, Va., 1769. Three children, among them, Olivier McClure, of Morgantown, W. Va.

WILLIAM McCLURE; who was president of the corporation of Dayton, Ohio, 1808. He was a Trustee of the First Presbyterian Church of Dayton, Ohio, organized 1801.

Miss Jean Wilkinson, Pueblo, Col., writes that she is a descendant of William McClure, who died in Tuscarora Valley, Pa. He had a son Willian, father of Harvey, father of Eleanor, mother of Miss Jean Wilkinson. William Sr. and Jr., were Revolutionary soldiers. She says, "Mollie McClure, the heroine of the Cherokee massacre, was of our family." See p. 156.

The Pennsylvania State Library gives records of the following Revolutionary soldiers, viz: Alexander, Andrew, Francis, George, James, John, Martin and Patrick McClure. Several of these are doubtless the same that appear in the Virginia records.

MISCELLANEOUS.

JOHN McCLURE (1725-1777), founder of the family in Botetourt county, was a son of Halbert and Agnes McClure of p. 135, and not a son of Arthur as stated on p. 142.

Besides the three sons, Samuel, Malcolm and Nathaniel, of whom sketches are given on p. 149; he had nine children, viz: Alexander, Mary, Agnes, Jennet, Hannah, Rebecca, Halbert, Moses and John.

John is said to have been born 1775, and was therefore the youngest. Married Isabella McCorkle. Nine children, viz: Samuel, Andrew, James, Capt. John A. who m. a Wilson, parents of Mrs. N. J. Baker, of Nace, Va.; William, Catherine who m. a Flaherty, Mary who m. a Kish, Joseph and Margaret. A granddaughter, Margaret, in Missouri. Samuel and William em. to Kansas.

Alexander McClure, b. 1797, who m. Sarah Hardy of Bedford County, Va., and em. to St. Louis, Mo., doubtless belongs to this family.

In Hening, Vol. 7, p. 181, James, John, James, Hugh and Halbert McClure, in Capt. Alexander Sayer's company, were paid Aug. 31, 1758, for military service. Alexander and Moses McClure, for provisions.

The Virginia State Library gives a fragmentary record of the following McClures in the Revolutionary War: Capt. David McClure, Capt. William McClure. First Lieutenants Andrew, Francis and John McClure. Ensign Geo. McClure. Privates Alexander, Andrew, James (Navy), John, Nathan, Patrick, Robert, Samuel and William McClure.

In the War of 1812: Privates Alexander, Andrew, Arthur and Samuel.

We find in the "Official Records, War of the Rebellion" that more than forty McClures served as officers in the Civil War. Among them—

From INDIANA (Wabash), Lieut. T. W. McClure.

From ILLINOIS, Col. John D. McClure, Capt. Geo. W.

MISCELLANEOUS.

McClure, Capt. T. J. McClure, Capt. Samuel M. P. McClure, Lieut. James A. McClure.

From IOWA, Lieut's H. M. McClure and Joseph D. McClure; Serg. J. M. McClure, who died in prison at Andersonville, Ga., Sept. 8, 1864.

From KANSAS, Capt. James R. McClure.

From KENTUCKY, P. McClure, died prison, Andersonville, Ga., May 10, 1864.

From MICHIGAN, R. McClure, died prison, Andersonville, Ga., Sept. 9, 1864.

From NEW YORK, Lieut. Thos. J. McClure.

From MISSOURI, Capt. T. J. McClure, Serg. Joseph McClure.

From OHIO, Capt. Addison S., son of Charles McClure, Maj. Daniel McClure, Lieut. Geo. D. McClure, Maj. John McClure, Capt. Oliver S. McClure, Capt. Wm. H. McClure, Capt. Wm. M. McClure.

From PENNSYLVANIA, Col. Alexander K. McClure, Capt. William McClure (15th Cavalry), Capt. Wm. M. McClure, (2nd Artillery), Dr. Samuel McClure.

From WISCONSIN, Capt. Wm. McClure.

Confederate soldiers:

GEORGIA, Lieut's J. J. McClure, (Clay Co.), and W. H. McClure, (Pike Co.).

MISSISSIPPI, James E. McClure.

TENNESSEE, Lieut. Robert G. McClure.

NATHAN McCLURE, of Russell County, Ky., was a member Kentucky Constitutional Convention, 1849. Member of the House, 1833-'39; Senate, 1848, '61-'63.

BRYAN S. McCLURE, Ky., Legislature 1871-'73.

R. C. McCLURE, Louisa, Ky., minority leader Ky. Legislature 1912.

JOHN D. McCLURE, Grand Master of the Grand Lodge of Ky., 1849. Grand High Priest, Ky., 1854.

REV. W. K. McCLURE, Methodist minister, 1914, Perryville, Ky.

The name occurs frequently in Missouri. Mrs. W. C.

Wilson, a descendant of John McClure, of Rockbridge Co., Va., states that Miss Nellie McClure, now Mrs. W. J. Harbicht, of Wentzville, Mo., is a niece of Joel and Milton McClure, who em. from Kentucky to St. Charles County, Mo., about 1825. There is a Margaret A. E. McLure Chapter, U. D. C., St. Louis. Mr. Claude McClure belongs to the family for whom the town, McClure, Ill., is named.

JOSIE McCLURE, b. 1864, in Gallatin, Mo., m. Wm. W. Martin.

REV. W. G. McCLURE was pastor, 1890, the Southern Methodist church, Marshall, Mo.

Miss Mary McClure, Madison, Ind., is said to belong to the Southern family.

APPENDIX.

THE ALEXANDER FAMILY.

This family came to Virginia from County Donegal, Ireland. REV. JAMES ALEXANDER was pastor of the Raphoe Presbyterian church, 1678-1704. ARCHIBALD ALEXANDER was an Elder im the Taboyn (now Monreagh) church about the same time.

The Alexander genealogy is given in Roger's Memorials of the Earl of Sterling and the House of Alexander, and Chart by Francis Thomas Anderson Junkin, LL. D., Chicago, from which the following is taken:

A Norse settlement was early established in Arran and Bute and other islands in the West of Scotland under the Viking Conn Chead Chath of the Hundred Battles. His descendant, Viking SOMERLED, about 1150, exercised powerful authority in the Western Isles, disputing the sovereignty of Scotland with David I. In 1164 he entered the Firth of Clyde with a fleet of one hundred and sixty vessels, intending to usurp the Scottish Crown. He was defeated at Renfrew and there slain. (Chron. Man. A. D., 1104-1164). He married, about 1140 (second wife) Affrica, daughter of Olave the Red, King of Man, and had three sons: DOUGAL, from whom came subsequently the Ducal House of Argyle; ANGUS, the third son, who became Lord of the Isle of Bute; and RANALD, the second son, who became Lord of the Isles of Mull, Kintyre, &c. His son DONALD was the father of ANGUS (d. about 1290), whose grandson JOHN, Lord of the Isles, married Margaret, daughter of Robert II, King of Scotland, grandson of King Robert I, the Bruce. Her descent from the old English kings of the House of Cerdic is as follows: King Ecgberht, d. 836; his son, King Ethelwulf, d. 855; son, King Alfred the Great, d. 899; son, King Edward the Elder, d. 927; son, King Edmund, d. 946; son, King Edgar, d. 975; son, King Etheldred the Unready, d. 1016; son, King Edmund Ironside, killed 1016; son, Edward the Confessor; daughter Saint Margaret, who m. 1068 Malcolm III, King of Scotland, d. 1093; son, King David I. of Scotland, d. 1158; son, Henry, Earl of Huntingdon; son, David, Earl of Huntingdon, brother of King William IV, the Lion; second daughter, Isabella, m. Robert Bruce, Lord of Annandale; son, Robert Bruce, the Claimant; son, Robert Bruce Earl of Carrick, who m. Isabella, Countess of Buchan, of the family of Macduff; son, Robert I, the Bruce, King of Scotland, b..1274 and d. June 7, 1329, m. Isabel of Mar; daughter Marjory m., about 1316, Walter Fitz Allan, the High Steward of Scotland; son, Robert II. King of Scotland, 1370, and founder of the Stewart (or Stuart) dynasty; his daughter Margaret m. JOHN, Lord of the Isles, the father of Alexander, Lord of Lochaber, whose son, MacAlexander, is looked upon as the real founder of the House of Alexander. His descendant, Thomas Alexander, in a legal instrument dated March 6, 1505, is mentioned as Baron of Menstrey. His son, Andrew Alexander, Baron of Menstrey, died prior to 1527. His wife was Katherine Graham. Their son, Alexander Alexander, Baron of Menstrey (1529), m. Lady Elizabeth Douglas, daughter of Thomas Douglas, eldest son of Sir Robert Douglas of Lochleven by his wife Margaret, daughter of David Balfour of Burleigh, an ancestor of the Earls of Morton. (See Douglas' Peerage, vol. II, p. 273). Alexander Alexander had a son, Andrew Alexander, Baron of Menstrey (1544), whose son, Alexander Alexander, Baron of Menstrey, d. about

1565. His wife was Elizabeth Forbes. His son, Wiliam Alexander, had two sons, Alexander Alexander, Baron of Menstrey, who was the father of William Alexander, Earl of Sterling, and Thomas Alexander, b. in Scotland 1630, but removed to Ireland 1652 for distaste of the Rump Parliament of Cromwell. "An intense Presbyterian, but loyal to the Catholic Stuarts, of whom he was a blood kinsman."

A daughter, MARGARET, m. Joseph Parks, who occupied lands in County Donegal. Daughter, Margaret. A son, WILLIAM, "remarkable for his carpulency, married and had four sons: Archibald, Peter, Robert and William.

A. ARCHIBALD, the eldest, was b. Cunningham Manor, County Donegal, Feb. 4, 1708, and m. Dec. 31, 1734, his first cousin, MARGARET PARKS, "a pious woman, of a spare frame, light hair and florid countenance." (Foote's Sketches). He did Colonial service; Captain in the Sandy Creek Expedition. Eight children.

I. ELIZA, b. in Ireland Oct. 1735. Came with her parents to Pennsylvania 1736, and Augusta County 1747, finally settled on Timber Ridge. She m., about 1754, John McClung of Rockbridge (b. 1732), whose sister MARY m. Judge Samuel McDowell, of Rockbridge. Three children:

1. Joseph, lived and died on Timber Ridge. Left descendants.

2. William, m. sister of Chief Justice Marshall. Was a distinguished Judge in Kentucky. Father of Col. Alexander K. McClung and Rev. John A. McClung, D. D. See "Marshall Family," Virginia.

3. Margaret, the oldest child, was b. October, 1755; "a famous beauty;" m., about 1775, ROBERT TATE. Died Sept. 23, 1839. Buried at Bethel. Son, James Tate; father of John Addison Tate; father of Margaret Letitia Tate; mother of Josie (Gilkeson) McClure. See Tate Family.

II. WILLIAM, the second child, was b. at Nottingham, Penn., 1738; married Agnes Reid; nine children, among them ANDREW ALEXANDER, 1768-1844, who m., 1803, Anne Dandridge Aylett (1778-1818), daughter of Col. William Aylett and Mary Macon, of King William County,

Va., and Rev. ARCHIBALD ALEXANDER, D. D., LL. D.,
(1772-1851), President of Hampden Sidney College and
founder (1812) of Princeton Seminary.

Capt. Archibald Alexander m., second, 1757, Jane McClure, daughter of James McClure, of Augusta County.
See p. 125.

B. PETER ALEXANDER d. in Londonderry. His wife
and children came to America, settling probably in Tennessee or Kentucky.

C. ROBERT ALEXANDER, a Master of Arts of Trinity
College, Dublin, came to Augusta 1743, and in 1748 established, near Old Providence Church, on land now owned
by Samuel Finly McClure, the Augusta Academy, "the
first classical school in the Valley," the beginning of Washington and Lee University. He m. Esther Beard, daughter of Thomas Beard, who d. in Augusta, 1769. He died
testate 1783. Ten children.

I. ELEANOR, m., June 26, 1790, Samuel Wilson.

II. SARAH, m., April 5, 1786, John Wilson.

III. PETER, m., March 27, 1787, Jennie, daughter of
Samuel Steele, of Augusta County. Em. to Woodford
County, Ky.

IV. ROBERT, m., Jan. 28, 1796, Jane Beard, daughter
Mary Martha, m. April 2, 1855, second wife, William C.
McClure.

V. William, m., Nov. 29, 1793, by Rev. John Brown,
Sarah Henry.

VI. Ann; VII. Esther. VIII. Hugh; IX. James; X.
Thomas.

D. WILLIAM ALEXANDER, wife Martha, died intestate
Augusta, 1755. Six children: William, who died 1768,
James, John, Agnes, Mary, George.

James Alexander, m. Sept. 11, 1759. He died testate
1809. Five sons and five daughters, viz: ANDREW, m.,
Sept. 16, 1788, Nancy, daughter of John Hamilton; GABRIEL; JAMES; JOHN, m., Nov. 27, 1788, Sarah Gibson;
WILLIAM; DORCAS, m., Feb. 19, 1794, Samuel Pilson, Jr.;

THE ALEXANDER FAMILY.

ELIZABETH, m., Feb. 1, 1785, Samuel Tate; MARY, m., 1785, Charles Campbell; MARGARET; MARTHA.

Closely related to these four brothers was JAMES ALEXANDER, member, 1740, of Tinkling Spring Church. He was doubtless the father or brother of—

A. ANDREW ALEXANDER, who d. testate in Augusta, 1789. Wife, Catherine. Sons, JAMES and ANDREW, Jr., who m., about 1778, Martha McClure. He d. 1787. Two children. See p. 25.

B. FRANCIS ALEXANDER, lived on Long Meadow. Wife, Elizabeth. Died testate 1792. Sons, GABRIEL; JOHN, m., Feb. 13, 1791, Rachel Miller; FRANCIS, "to be schooled by Gabriel;" WILLIAM; JANNET, m., March 11, 1790, her first cousin, John Alexander; DORCAS, m., 1798, Aug. Smith.

C. GABRIEL ALEXANDER settled, 1749, on South River. Wife, DORCAS. Died testate 1779. Six children, viz: FRANCIS, m., Dec. 29, 1790, Elizabeth McClure (See p. 25); GABRIEL; JAMES; JOHN, m., March 11, 1790, Jannet Alexander; DORCAS LACKEY; MARGARET.

AGNES, wife of James McClure, founder of the Augusta County family, was probably an Alexander.

THE BAXTER FAMILY.

In Foote's Sketches of Virginia, Second Series, p. 262, we read, "George Baxter and Mary Love were emigrants from Ireland at a very early age, landing on the banks of the Deleware. The parents of George dying soon after their arrival, he was received into the family of Thomas Rodgers. This gentleman had married Elizabeth Baxter and emigrated from Londonderry to Boston, Mass., in 1721. In about seven years he removed to Philadelphia. George Baxter, when of mature years, followed his emigrating countrymen in their search for a home on the frontiers of Virginia and chose his residence in Mossy Creek congregation, once a part of the Triple Forks, and afterwards of Augusta church. In the course of his life he represented his county in the Legislature about fifteen times. He reared his family according to the customs of his fatherland and the habit of his emigrating countrymen, in industry and economy, giving all an English education in a manner as liberal as circumstances would permit, choosing if possible, one child of talent for a liberal education and a professional life.

MARY LOVE, his wife, left among her descendants a memory precious for her exemplary piety and prudent conduct as a wife and mother, in situations calling every day for the exercise of Christian graces and seldom offering occasion for the lofty display of any accomplishment. The lives of her eight children were her best eulogy. Vigor, frankness, uprightness and industry characterized all the members of the family, reared in the simplicity and hardships of a frontier life. The mother laid the foundation of morals and religion in her children while they were young, and expressed the most decided unwillingness to part with any of them till their faith in Christ was established. Her unremitting attention to the spiritual concerns of her children was followed by the unspeakable reward of seeing them all consistent professors of religion, according to the faith she trusted for her own salvation. The Bible, the Sabbath, the Assembly's Catechism, the preaching of the gospel, family worship and private instruction, were things of solemn interest to the family from the earliest recollections, and connected indissolubly with the memory of their parents, the influence was tender and perpetual. The image of the mother stood before the children rejoicing when their triumphed, and weeping when they sinned."

His wife was a dau. of Col. Ephraim and Elizabeth Love. He was for many years an elder in the Mossy Creek church. A Revolutionary soldier; qualified lieutenant Nov. 23, 1778, and captain March 12, 1779. Of their eight children, Rev. George Addison Baxter, D. D., the second son and third child, was born July 22, 1771. Mar. Ann Fleming, dau. of Col. Wm. Fleming, of Botetourt County. Was President of Washington College (now W. and L. U.), and professor in Union Theological Seminary.

Rebecca Baxter, the youngest child, was b. 1783, and d. Feb. 25, 1817. Her grave is marked in Bethel Cemetery. She m. James Tate, great grandfather of Josie Charlton Gilkeson, who m. James A. McClure.

THE BUMGARDNER FAMILY.

This family belongs to Switzerland, where RUDOLPH BAUMGARTNER, of Basle, was one of the leaders in the rebellion that won freedom from Austria. A large portrait of LUCAS BAUMGARTNER in armor is shown in one of the art galleries of Munich.

HANS BAUMGARTNER, the founder of the Augusta family, settled on "Stony Lick, a branch of the Shanandore, opposite Great Island." This is in the present limits of Rockingham County. His deed for 400 acres of land is dated September 25, 1746. His will was proven March 22, 1751, (See Chalkley III, p. 21). Sons, John and Christian; daughters, Mary, Elizabeth and Madley. grandson, Jacob Burner.

The family have no record of JOHN, the older brother. The name, however, occurs several times in the records of the Valley.

Godfrey Baumgardner settled on New River, 1762.

Rudolph Bumgardner was living in Hampshire County, 1784, married and had six children.

Christian, David and Peter Bumgardner were living single, 1785, in Shenandoah County.

CHRISTIAN BUMGARDNER, the younger brother, b. about 1740, settled on a farm adjoining the present Bumgardner home, Augusta County. Chalkley, vol. II, p. 49, shows that he was a Colonial soldier, serving with Washington in his campaign of 1754, for which he was pensioned. A letter from his friend, Gen. Daniel Morgan, recently lost or mislaid, shows that he was also with Washington at Braddock's defeat. He qualified Lieutenant of Foot Nov. 17, 1757. He did service in the Revolution and died the day following his return from the war. His son,

THE BUMGARDNER FAMILY.

I. JACOB BUMGARDNER, b. Feb. 8, 1769, and died Aug. 25th, 1859. He m., June 28, 1785, MARY Waddle, daughter of John and Mary Waddle, a family from Saxe-Weimer, Switzerland, who owned the present Bumgardner home and who gave the land for Bethel Church. Ten children:

1. Christian, b. 1786. Em. to Kentucky. A son, Anthony Wayne.
2. Polly, m. Jacob Kunkle, of Augusta.
3. Jacob, ensign to Capt. Samuel Steele, War 1812. Emigrated to Kentucky. Son, Jacob.
4. David; d. s.
5. William, Em. to Kentucky.
6. John, m. Jane Clarke, of Staunton.
7. Betty, m. Abel Gibbons. Lived near Bethel.
8. Sarah, m. Alexander McGilvray. Died in Greenville. Twelve children, the late Rev. William McGilvray, of Richmond, and Sarah, wife of J. Alexander Bumgarder, of Bethel.
9. James, m. Malinda McCorkle, of Rockbridge County. He is mentioned in Waddell, p. 433. "A meeting was held at Greenville June 11, 1836, to attempt to raise a company of militia to engage in war against the Creek Indians, which was called to order by James Bumgardner."

Five children—

(1). Col. William Bumgardner, m. Pocahontas Happer. Three children.

(2). Jacob Alexander Bumgardner, m. Sarah McGilvray. Five children living, viz: Eugenia, single; Malinda, m. Charles Berkeley and has four children; Edwin, m. Janie T. McClure; Tillie, m. James M. Lilley; Alexander.

(3). Mary, m., Dec. 27, 1863, Capt. James Bumgardner. Six children living.

(4). Eugenia, m. Archibald Alexander Sproul.

(5). Betty, m. the late Livingston Murphy, M. D., for many years the distinguished Superintendent of the Morgantown Asylum, Morgantown, N. C. Three children, viz: Mary, Dr. Alexander Murphy, of Goldsboro, N. C., and Dr. James Murphy, of New York.

THE BUMGARDNER FAMILY.

10. Lewis, b. Aug. 17, 1806, and d. Oct. 11, 1890. He m. Nov. 5, 1833, in Lexington, Ky., Hettie Anne Halstead (Nov. 10, 1815—Jan. 25, 1872). He lived for a number of years in Carroll County, Mo., where his older children were born. Returning to Virginia, he was a merchant in Greenville and later in Staunton. Ten children:

(1). Capt. James Bumgardner, b. Feb. 18, 1835, graduate University of Virginia, Captain Co. F., 52nd Regiment, C. S. A.; a distinguished lawyer, Staunton, Va.; m. Mary Bumgardner Dec. 27th, 1863. Six children living, viz:

James Lewis Bumgardner, a graduate of W. and L. University, lawyer, Beckley, W. Va,; m. Ophia Ellison. Two children, Mary Mildred, Eunice Ellison.

Rudolph, graduate W. and L. University; lawyer, Staunton, Va.; m. Nannie, daughter of Rev. W. N. Scott, D. D., of Staunton. Two children, Mary Margaret, Rudolph, Jr.

Four sisters, single, viz: Minnie, Gussie, Eugenia, Nellie.

(2). Mary, b. Aug. 9, 1836; m. Andrew Wellington McClure, Sr. (q. v.)

(3). Daniel Halstead, Aug, 27, 1838—July 23, 1847.

(4). William, Aug. 29, 1840—Oct. 2, 1841.

(5). Sarah Catherine, b. March 18, 1842; m. M. T. McClure, Sr.

(6). Augusta Virginia, b. Feb. 7, 1844; m., Dec. 9, 1897, J. F. Tannehill, Sr.

(7). Lewis, Jan. 24, 1846—May 18, 1847.

(8.) Jacob, b. April 1, 1848, and d. May 16, 1902. He m., March 21, 1894, Minnie May Jones, of Allentown, Pa. Two sons, Walter and Jacob.

(9). Malinda, Aug. 20, 1851—Nov. 20, 1854.

(10). Lewis Milton, b. Nov. 22, 1853; graduate U. of Va.; lawyer, Staunton, Va. Died single May 6, 1888.

THE HALSTEAD FAMILY, English, settled early in New York. John Halstead was captured the night the English took the city, Revolutionary War, and like many other prisoners, died of suffocation. His wife was a Nichols, a descendant of Gov. Nichols of New Jersey. Three sons, Christopher and John, who lived and died in New York

City. The office of the Corporation of Trinity Church contains a number of Halstead records; marriages, births, baptisms, etc. Daniel, who was five years old when his father died, married, in Trinity Church, Hetty Sprong and em. to Lexington, Ky. Nine children, viz: John, married and lived in Kentucky; Christopher married and lived in Illinois. His copy of Josephus is now owned by Rev. J. A. McClure. Maria, d. s.; Sarah, m. James Harvey Burch, of Missouri; Eliza, m. James Scantland, of Kentucky; James married and lived in Illinois; Alexander lived in Illinois; Hettie Anne m. Lewis Bumgardner, of Virginia; Mary Ellen m. Zophar Case, of Cleveland, O., parents of Warren Case, who m. Linda McClure.

THE SPRONG FAMILY traces its descent back through the Brower and Bogardus families to Aneke Jans, granddaughter of William of Orange and the Bourbon princess Charlotte. Aneke Jans settled in N. Y. with her husband, Roeloff Jansen, in 1630. She m. second, Everardus Bogardus, the second pastor of the church in New Amsterdam.

Bogardus died December 27, 1647, and Aneke Jans in 1663. They owned seventy acres in what is now the heart of New York City. The subsequent litigation growing out of this and the lease by the Trinity Corporation, is generally known. (See Bogardus Chart, Genealogical Department, New York City Public Library).

The SPROUL FAMILY, of Augusta County, is probably descended from JOHN SPROUL, an elder in the Raphoe Presbyterian church, Co. Donegal, Ireland, 1700.

WILLIAM SPROUL settled on Moffett's Creek about 1750, and died testate 1806. He was twice married; first, Aug., 1757. Wife, JANE. Four children, viz: JAMES, ALEXANDER, who m. May 15, 1781, Jane Beard; WILLIAM, MARGARET, who m. a McCutchen.

He m. second, June 23, 1773. Wife, SUSANNA. Eleven children, viz: JOSEPH, OLIVER, JOHN, CHARLES, who m. Margaret, a daughter of Dr. Alexander Humphreys, of Staunton, was a lawyer in Frankfort, Ky. SAMUEL, JANE who m. June 23, 1793, John Weir, of Rockbridge. SIDNEY, who m. Jan. 21, 1799, Joseph Beard. MARY, d. s., MARTHA m. Robert Hutchinson, of Rockbridge, FANNY, who m. Thomas Thompson, of Augusta. NEMSY.

JOHN SPROUL, b. about 1785, m. about 1820 Matilda Scott, daughter of the Revolutionary War pastor of Bethel church. He was ordained an elder in Bethel Sept. 18, 1831, and died May 22, 1849. Seven children.

(1). Susan Jane, b. July 7, 1822, m. Samuel Bell.

(2). Frances Elizabeth, b. February 15, 1824, m. Wm. White, of Lexington, Va.

(3). Emeline, born March 1, 1826.

(4). William Scott, b. February 4, 1828, d. s.

(6). Matilda, born Jan. 26, 1833, d. s.

(7). Martha Ann, b. February 20, 1837, m. Archibald G. Christian, parents of Lee Christian, who m. Julia Smith, and others.

(5). Archibald Alexander, b. April 29, 1831. Ordained an elder in Bethel August 4, 1866. He married Eugenia Bumgardner. Nine children, viz:

James Bumgardner, b. Sept. 7, 1860, d. i.

John Alexander, b. Jan. 9, 1862, Clifton Forge, Va.

Wm. White, b. April 17, 1863, an elder in Bethel. Mar. Kate Lapsley, of Anniston, Alabama. Four children.

Matilda Scott, b. February 9, 1865, d. s.

Malinda, b. Sept. 15, 1866, m. John Marshall McClure. Died.

Archibald, b. January 10, 1868, a distinguished graduate of W. and L. U. Married Mary Cotton, of New York. Died. A son.

Eugene E., born August 3, 1870, d. s.

Frances, living in Clifton Forge, Va.

Hugh Bell, born November 12, 1873, student W. and L. U. Married Agnes Miller, of Staunton. Five children.

THE MITCHEL FAMILY.

ELEANOR MITCHEL, a widow, and her son, JOHN MITCHEL, settled in Augusta County, 1747. Their farm cornered John Tate and David Doak.

JOHN MITCHEL died testate 1771. See Chalkley III, p. 120. Wife, Elizabeth. Six children.

I. THOMAS. His Bible is now owned by M. T. McClure, Sr., a desendant. He was born Sept. 23, 1732, and died testate Dec. 30, 1806. Revolutionary soldier. He m. first, Dec. 27, 1757, Elizabeth McClanahan Moor, widow of Samuel, son of Andrew, who was accidently killed 1752. A son, Andrew, b. 1750 and died August 10, 1791. Wife, Martha, parents of Mrs. Andrew Lusk.

Elizabeth Moor; d. August 1, 1777, "in the forty-ninth year of her age." Five children, viz:

1. William, b. October 5, 1761, m. September 15, 1785, Agnes Brownlee.

2. Thomas, Jr., b. Dec. 7, 1763. Went West.

3. Elizabeth, b. March 1, 1766, d. Sept. 11, 1850. M. September 28, 1810, James Fulton (1755-1834). Buried at Old Providence. John McClure, a nephew, lived with them and inherited from them ,1819, the farm now known as the McClure Homestead, one mile north of Old Providence church.

4. Mary, b. December 23, 1768, and died 1795. She m. January 15, 1789, Andrew McClure.

5. Isabella, b. Sept. 1, 1771, m. Dec. 16, 1807, John Doak. Buried at Bethel.

Thomas Mitchel m. 2nd Nov. 6, 1781, Elizabeth Wales. She d. July 11, 1806. Dau.,

6. Jane, b. Dec. 1, 1782, m. 1st, James Mateer, son of James Mateer, Sr., and Elizabeth Wright. James Mateer d. 1812, testate. Four children, viz: William, Mitchel, Eliza, Isabella. Jane Mitchel Mateer m. 2nd, William Mateer, a widower, with five children, viz: Catherine, John, Polly, William, Jr. and Virginia Mitchel, b. Sept. 28, 1824, baptized at Bethel April 1, 1825. She m. a Wood in Missouri. They em. to Missouri, and had three children, viz: Ann who m. a Miller, Sally who m. Peyton,

and Samuel, father of Mr. Albert Mateer living (1913) Calwood, Mo.

In a letter from Callaway County, Mo., June 24, 1830, I read that Isabella Mateer was taken sick on her way out and that "Mr. Hannah took a carriage and brought her home." One would judge from the Augusta records that the Mateers, Mitchels, Doaks and Hannahs were related.

The following letters from William Mateer, son of James Mateer, and Jane Mitchel, to his first cousin, John McClure, are of interest:

"Caloway County, Mo., Dec. 20, 1827.

Dear Cousin,

I take my pen in hand to let you know that I am well and hope that these lines may find you and family well. In the first place would let you know that I had a tedious trip to this country. I was two months and four days from time I left home until landed on Salt river, and found this country much what I expected to see in every shape and form; found a great quantity of good land and a vast quantity of perrare, and some poor looking land. This country lays very level, but still roling enough to carry the water off the ground in a short time. I have been up the Misssippie 147 miles from St. Louis, and from there 75 or 80 miles up toward the headwaters of the Salt river, and from there across to the Missouri river. I have likewise been 200 miles up the Missouri from mouth and find the country much the same. Springs is very scarce, but I find the people that make use of creek water equally as healthy as them that have springs, and considered by some more so, but water can be had convenyently by sinking wells from ten to thirty feet. Stilling is a good business in country where a man has a mill of his own to grind his grain. Mills are scarce in this country; almost all horse mills, and then you must grind yourself and with your own horses and give the sixth bushel. You can get stills in country from 20 to 25 per cent lower in the gallon than in Virginia, and the head throwed in. Whiskey is two shillings per gallon by the barrel and from 50 to 62½ by the retail, and a great deal of it is sold in that way. Corn is one dollar per barrel, pork $2, beef $1.50, wheat 37½, rye two shillings, coffee from 20 to 25 cents, sugar 12½ cents, and store goods is as cheap as in Virginia; all but iron, that is ten cents, though but little of it made use of. Horses goe without shoes, and a man that has to get his plow-irons sharped twist a year thinks his blacksmith work a heavy tax on him.

There can be no good entrys of land in this neighborhood; there can be a great settlement made about 6 miles from Henderson's on the headwaters of ——— River. There is no settlement within several miles, so a man may situate himself just as he sees proper. If you come to this country next fall, come through Kentucky, cross the Ohio at Albany 4 miles below the falls and from there to Vinsane on the Wawbash, which is 120 miles, and from there to St. Louis 160 m., to St. Charles 20 m., from there here 85 m. The distance from where you live to St. Louis don't excede 770 miles that rout. Right me a letter and let me know your notion sertenly about coming to this country by the first of Aprile. I want you to fetch me a Virginia wife out here; some hansome and clever girl. Tell Betsy young girls is ready sale here; but old maids rate at 25 cents a hundred. I have not room to write any more; give my compliments to uncle and aunt, Jane and Betsy. Likewise take them to yourself, and believe me your sincere friend,

WILLIAM MATEER.

Mr. John McClure, Greenville, Va."

John McClure, since the last letter, had made a trip to Missouri on horseback, with Matthew Pilson, his brother-in-law.

"Ralls County, Mo., December 20, 1829.

Dear Cousin,

I take this opportunity of informing you that I am well and hope that these lines may find you and your family well I have nothing to write worth your attention. David, Hannah and Robert got to Salt river and all is well except Jane Henderson, and she is able to walk about, but mends slow. I see no change in times since you left here, only the people still continue coming into this section. I had forgotten when I said there had been no change or alteration since you left here; there has been one case of murder in New London a few days ago. We in the upper part of the county disown London altogether and wish them great success in killing one another until the place becomes properly clensed. I believe I have nothing more about our country or people. I would be glad to hear from you and know what part of Missouri you and Mr. Pilson is best pleased with and whether Mathew feels like moving since he has got home or running on a while longer. Write to me without delay. Give our respects to uncle and Betsy McClure, Mathew Pilson and Jane. Nothing more.

Your humble servant,
WM. MATEER.
Mr. John McClure, Greenville, Va."

II. Robert Mitchel, b. about 1734. Surveyor, 1774.

III. John Mitchel, d. 1783.

IV. James Mitchel, b. about 1740. Revolutionary Soldier, Lieutenant in Captain James Tate's company. Was a charter member and with Colonel Robert Doak, was one of the first Elders in Bethel Church at its organization 1779. A number of his books are now in the Library of the Author. Among them the Trial and Triumph of Faith, inscribes "James Mitchel, his hand and pen, June ye 12 1773." He died 1806. Four children:

1. Thomas, m. April 4, 1786, Margaret, dau. of James Callison, Died 1816. Seven children, among them, Wilson Mitchel, b. Jan. 10, 1796.

2. James, m. dau. of William and Jane Brown.

3. Sally, m. May 30, 1785, Robert Beard.

4. Elizabeth, m. Feb. 16, 1793, Robert Callison.

V. Eleanor, m. Mathew Willson, Sr. an Elder in Bethel. Eight children, among them Mathew Willson, Jr. also an Elder in Bethel.

VI. Mary, m. a Wright.

JAMES CALLISON, "otherwise lately called James Callison in the settlement, Albemarle Co. Va.," and his wife Isabella, settled in Augusta Co. 1749. He died testate 1789. Eleven children: John, whose dau. Isabella m. Feb. 25, 1801, Joseph Evans. James, em. to Ky., sons Anthony and Isaac, of Bath Co. Va. Margaret, m. 1786, Thomas Mitchel. Robert m. 1793, Elizabeth Mitchel. William, Dorathy, Jean, Agnes, Mary, Eleanor, and Isabella.

James Scott Callison, who m. Carrie Pilson McClure, and his sister, Mary Dell, who m. Rev. C. D. Waller, son and dau. of the late James Callison, doubtless belong to this family.

THE McCOWN FAMILY.

JOHN McCOWN, founder, settled in Rockbridge County about 1740, militia service 1742; Constable in Forks of James, 1746, died testate, 1783. His farm 437 acres deeded by Benjamin Border 1750, is still in the family, the home of Robert McCown near Rockbridge Baths. His son JOHN McCOWN, Jr., was born about 1740, father of Capt. John McCown, b. March 17, 1784. He m. Polly Culton, (May 3, 1786, June 23, 1869.)

"Died on the 11th, of April 1850, at his residence in Rockbridge Co., after a few days of severe suffering from a violent attack of pneumonia, Capt. John McCown, in the 66th year of his age. In the death of our friend, society has lost a most valuable citizen, and the Presbyterian Church, of which he was a member, and a ruling elder, one of its chief pillars. Whilst he was kind and open hearted to other denominations of Christians, he was a whole-souled Presbyterian; devoted to the standards and policy of the Presbyterian church. The cause of Zion seemed to be always near his heart, and whenever a call for aid for any pious and Christian enterprise was presented, his hand and his heart were always open.

But whilst we feel deeply the loss of so valuable a friend

and member of our church, we are consoled with the consideration, that having finished his labors in his Masters vineyard here, he has been taken up to receive his crown amongst the redeemed above. (Signed) D."

His son John Kinnear McCown was b. Feb. 24, 1811. He m. Mary Wilson. Issue, Robert, Horatio, Rev. James H., Agnes, Sarah who m. Samuel Wilson, Emma who m. Capt. John McNeil and Samuel Walter who m. Anne Halstead McClure.

"At a meeting of the Session of New Monmouth church, January 17th, the following tribute to the memory of elder John K. McCown was adopted and ordered to be spread on the minutes:

John K. McCown, the venerable elder of this church, died January 5th, 1892, in the 81st year of his life. Descended from a pious ancestry, this good brother confessed Christ in early life, and by the grace of God, held fast the profession of his faith unto the end. About fifty-two years he served the church as a ruling elder, having been ordained to that office in the year 1840, As a session, and in behalf of all our members, we would give expression to our high appreciation of his Christian character and official fidelity, and our sense of loss by his removal. His love of truth and righteousness, his devotion to the cause of God, his strong sense of duty, and his great decision of charter, made him a most valuable ruler in the house of God, while his influence for good could not fail to be felt by the whole community. Not only did he train his own house in the fear of the Lord, giving two of his sons to the ministry of the gospel, but his daily life was a constant rebuke to all evil doers. While we mourn his death, surely we may comfort ourselves with the word of the Lord and say 'Blessed are the dead which die in the Lord.'

By order of Session.

D. A. PENICK, Moderator."

THE PILSON FAMILY.

SAMUEL PILSON, founder of the family in Augusta, is mentioned by Foote in connection with the organization of Tinkling Spring Church, 1740. Two sons:

A. RICHARD PILSON; was living in Albemarle Co. 1773. His son ROBERT PILSON, m. AGNES McCLURE. Their children, Hugh, Ann, Richard and Polly, were living in Ohio, 1833.

B. SAMUEL PILSON, b. 1739, died after 1807. Road surveyor, 1773. Probably m. a Hutcheson. Three sons.

I. John, d. s. Elder in Bethel.

II. Samuel, m February 19, 1795, Dorcas, dau. of James Alexander. Em. to Ohio.

III. George, (1765-1833), m. Jan. 4, 1796, Elizabeth Thompson, b. 1764, d. December 17, 1861. Five children:

1. Jane, (June 14, 1797 - September 18, 1882). Mar. John McClure.

2. Phoebe, m. John S. Thompson.

3. Samuel, 1807-1811. Buried at Tinkling Spring.

4. Mary, m. Jacob Lightner. Two sons: Geo. P., living near Spottswood, Va., and the late Samuel A.

5. Matthew, m. November 5, 1834 Lavinia Finley. Five children left descendants. Others died single.

(1) Samuel Finley, who d. March 10, 1914. M. Ellen Finley, of N. C. Two sons:
The late Rev. Matthew Finley Pilson, and Edward Pilson, who died 1914.

(2) George, m. Ellen Lambert. Two sons, Lacy and Blair Pilson.

(3) Dr. William Pilson, son, W. H. F. Pilson, Lawyer, Staunton, Va.

(4) Frank, m. Carrie Finley of N. C. Two daughters.

(5) Matthew Thompson, m. Anna Hogshead; five children.

Matthew Thompson settled in Augusta County about 1740. Died in 1753. His will is recorded Staunton, Va., mentioning sons, William, John, Matthew, and Matthew his grand son. This family while doubtless related is not the one directly connected with the McClures through the Pilsons.

Two brothers, Charles and William Thompson came to Augusta Co. from Penn. prior to 1750. Charles' name appears only once on the county records, 1752. In the Americanized Encyclopædia Britannica Vol. X p. 6835 it is stated that he was born in Maghera, Co. Antrim, Ireland, Nov. 29, 1729. He came to America in 1740; studied in New London Penn. and later conducted a Quaker school at New Castle. In September 1774, he went to Philadelphia with his bride, a sister of Benjamin Harrison, the Signer, having been chosen Secretary of the first Continental Congress. He died in Lower Merion, Penn., Aug. 16, 1824. He declined a place in President Washington's Cabinet that he might complete his translation of the Scriptures, a set of which he presented to each of his nephews and nieces. The books are still in the Pilson family.

They were joined by a third brother, MATTHEW, their father accompanying him died envoyage and was buried at sea.

WILLIAM m; 1761 d. 1815. Nine children:
I. Mathew. who d, 1806.
II. Margaret.
III. Elizabeth, m. a Wilson.
IV. William, (1770-1835). Buried at Bethel.
V. Rachel, m. April 5, 1780, Alexander Berryhill.
VI. Martha, m. March 29, 1787, Robert Talbert.
VII. Jane, m. Aug. 21, 1792, Thomas Brown.
VIII. John, m. Dec. 3, 1793, Jane Blackwood, (Nov. 5, 1768, May 28, 1842). Sons William and John.
IX. Mary, m. June 8, 1799, Robert Willson, Jr.

MATTHEW, m. Sept. 20, 1763. Wife SALLY, probably an Alexander. Died 1822. Four children:
I. Elizabeth, (1764-1861), m. George Pilson.
II. Jean, m. Dec. 27, 1791, Andrew Hunter.

III. Mary, m. Oct. 10, 1797, William Shields.
IV. Matthew, Jr., d. s.

THE DRAPER AND INGLES FAMILY.

GEORGE DRAPER m. ELEANOR Hardin in County Donegal, Ireland, came to Philadelphia 1728, and to Virginia about 1740. Constable 1747, died 1748, probably killed by Indians. Son, John, b. 1730; dau. Mary, b. 1732.

COL. THOS. INGLES, possibly a son of William Ingles, an Elder 1700, in Monreagh Presbyterian church, County Donegal, came to Virginia about 1740, settling 1749 in what is now Montgomery county. His son, COL. WILLIAM INGLES, was b. 1729 and m. 1750 MARY DRAPER above. He was commissioned Captain of Foote Aug. 24, 1754; Justice 1769; died testate 1782. His home was destroyed by Indians July 30, 1755, and his wife and two children carried away captive. For the full account see Foote's Sketches of Virginia 2nd, p. 149; Hale's Trans-Alleghany Pioneers, p. 11, and Waddell's Annals of Augusta County.

"Being a woman of extraordinary courage and tact, she ingratiated herself with the savages, making shirts for them and gaining their good will in a hundred ways. Her two older children were, however, separated from her, and she then determined to escape if possible. The narrative of her courage and sufferings on her trip home is almost incredible. She was absent about five months, of which time forty-two days were passed on her return."—Waddell, p. 144. She d. at Ingles' Ferry Va., February, 1815.

Six children:
I. Thomas, b. 1751, m. Ellen Grills.
II. George, died in captivity.
III. Susan, m. Gen. Abraham Trigg.
IV. Rhoda, m. Capt. Bird Smith.
V. Mary, m. John Grills.
VI. John, b. 1766, lived in Montgomery Co. Elder in the Presbyterian church. Mar. Margaret Crockett, of Wythe County. Nine children, among them Malinda

Charlton, b. 1805, and Margaret Crockett Ingles (1808-1878), who m. 1st, Thos. Hyde; 2nd, Wm. J. Gilkeson, grandfather of Josie Charlton Gilkeson, wife of James A. McClure.

THE GILKESON FAMILY.

WILLIAM GILKESON from the north of Ireland, settled in Lancaster County, Pa., about 1730. He m. Margaret Lynn, dau. of Hugh Lynn from near Philadelphia. He em. to Frederick County, Va., settling near Kernstown, 1765, where he died 1778. Buried at Opequon church. His will is recorded at Winchester, Va. Eleven children:

I. HUGH, born in Pa. 1748 and d. in Augusta County 1806. His will is recorded at Staunton, Va. He m. Elizabeth Guthrie (1746-1830), dau. of John and Ann Guthrie, who came to America on the same boat with the Gilkesons. Six children:

1. Margaret, (1773-1816), m. May 15, 1794, John Guthrie, Augusta County. Ten children.

2. Ann, m. Jan. 29, 1801, James Craig, Kentucky.

3. Elizabeth, b. Jan. 25, 1778, m. her cousin, David, son of Wm. Gilkeson.

4. DAVID (1782-1866), m. Polly Humphreys. Elder in Tinkling Spring. Six children, viz:

(1). Hugh Lynn, b. 1810, d. s. in Illinois, 1836.

(2). Margaret, 1812-1848, m. (2nd wife) 1839, Andrew Patterson, of Brownsbury, Va. Three children.

(3). David Carlisle, b. 1815 and died Aug. 22, 1864. Confederate soldier. M. Jan. 20, 1842, Harriet Newall Finley (1821-1886). Four children, viz., Virginia Ruth, Carrie Belle, m. 2nd wife, James W. Wallace; Samuel Finley, an Elder in Bethel, d. s. 1913; Elizabeth, who m. Samuel Brown, an Elder in Hebron.

(4). Elizabeth, b. 1820 and m. 1852, 2nd wife, Rev. W. W. Trimble, of Missouri.

(5). James, lived in Fauquier Co. Six children, viz., Betty, Mary, Carrie Belle, who m. a Guthrie of Miss.;

THE GILKESON FAMILY.

Carlisle, Harry and Samuel, who m. a Martin and lives (an Elder) in the Shemeriah congregation, Augusta County.

(6). John A., m. Dec. 18, 1851, Isabella Sterrett Humphreys, b. Jan. 17, 1831. Five children:

a. Hugh Finley, b. Oct. 18, 1852, m. Dec. 8, 1881, Lula Tate Larew, dau. of John T. Larew. Dau. Margaret.

b. Frank Humphreys, b. Nov. 18, 1854, m. Nov. 19, 1885, Mattie B. Hanger.

c. Rev. Charles David Gilkeson, b. June 30, 1863, m. Oct. 15, 1891, Margaret Leyburn, of Lexington, Va. Son,

(a). Charles Leyburn, b, April 26, 1899.

d. William J.   e. Emma J.

5. WILLIAM (1784-1864), m. Sarah, dau. of his uncle, John Gilkeson, of Frederick County, Va. Four children:

(1). Andrew Todd Gilkeson, m. Emma Heiskell. Four children, viz., William, Belle, Clara and Harry. The latter m. Mary McKee, of Buena Vista, Va.

(2). David Gilkeson, m. Mollie Gamble, of Ga. Three children, Lillie who m. Walter Guthrie, of Miss. Lula and Belle now (1914) living in Waynesboro, Va.

(3). Hugh William Gilkeson, d. 1900 in Ga.

(4). John Gilkeson, m. Bettie Patterson, d. without heirs, 1841.

6. JAMES, m. Eliza Crawford. Lived near Springfield, Ky. Two children:

(1). Elizabeth, m. Rev. A. A. Hogue, Lebanon, Ky.

(2). James A., m. a Hopper, sister of Rev. Jas. Hopper, of Perryville, Ky. Four children, Joseph, Maggie, James R., Katie.

II. WILLIAM, b. in Penn. Aug. 29, 1750, came with his parents to Frederick Co., 1765, thence to Augusta. The Staunton records give his deed dated May 1, 1780. "David Kerr to William Gilkeson of Frederick County, part of tract that Andrew Cowan formerly lived on in Beverley Manor." This farm, "Hillside," is still the Gilkeson home, now owned by the family of the late M. F. Gilkeson, his grandson. A record 1785 says he "was of good character here and below in Pennsylvania." Foote, 2nd p.

THE GILKESON FAMILY.

356, states that Rev. Conrad Speece preached at his house April 5, 1813. He m. SARAH LOVE (Aug. 29, 1752—June 27, 1826), widow of James Guthrie. He d. July 3, 1828. Their graves are marked at Bethel. Children:

1. Nancy, m. an Irvine. Buried at Bethel.
2. Margaret, m. Aug. 21, 1799, Elijah McClenachan.
3. Lovie, m. Dr. John Tate. Illinois.
4. Jane. Dismissed from Bethel church April 21, 1821. Is said to have m. a Herring.
5. William J., b. 1789 and m. Margaret Crockett Ingles, of Wythe Co., widow of Thomas Hyde. She left one son, Thomas Hyde, father of the family now living near Stuart's Draft, Augusta Co.

Seven Gilkeson children:

(1) Francis McFarland, b. Aug. 42, 1838. Is living Culpeper, Va.; Alumnus Washington College; mar. Fanny Greene. Eight children, viz: Fanny who m. Leon Nalle, Lovie B. who m. F. E. Porter, Frank, Wm. Irvine, Sadie, Mary who m. J. D. Coleman, J. Archibald, m. and lives in Raleigh, N. C., Nannie, m. H. V. Frazier.

(2) John William, b. March 2, 1840, baptized May 3, 1840. Died of pneumonia March 25, 1907. Farmer. Alumnus Washington College (1859), now Washington and Lee University. One of the first deacons of Bethel church and for many years Treasurer. First Lieut., 25th Va. Regiment, C. S. A. Was captured and confined for eighteen months on Johnson Island. He m. Margaret Letitia Tate. Seven children:

a. John Hansford, b. Aug. 4, 1869. Graduate and instructor Virginia Military Institute. Student U. of Va. Farmer.

b. Margaret Randolph, b. Oct. 12, 1870. Graduate Mary Baldwin Seminary.

c. Edna Ingles, b. Dec. 25, 1876. Graduate M. B. S. Teaching 1914 in Cairo, Egypt.

d. Anna, d. eight years old of diphtheria.

e. Josie Charlton, b. March 28, 1880. Graduate M. B. S. Mar. Dec. 31, 1903, James Alexander McClure. Six children: the youngest, Josephine Charlton, was b. Petersburg, Va., Sep. 13, 1914.

f. Rebecca Baxter, b. Nov. 22, 1882. A. B., Mary Baldwin Seminary, Ph. B., M. A., University of Chicago. Teacher (1914) Ward-Belmont College, Nashville, Tenn.

g. Mary Tate, b. Now. 5, 1889. Graduate Stonewall Jackson Institute; mar. Oct. 15, 1912, Wm. Allen Wallace.

(3). Sarah Margaret, b. Jan. 27, 1842. Living Staunton, Va.

(4). Irvine Waddell, b. March 12, 1844, died ———. Physician, Mint Spring, Va. Mar. Catherine J. Gilliam. Two children, Hansford, single; Catherine, who m. Dr. Wm. Deekens of Staunton, Va. Dau. Amelia Catherine.

(5). Lovie Jane Herring, baptized Apr. 23, 1848. Mar. Dr. Carter Berkeley of Staunton, Va. Six children, viz: Edmund, Charles Carter, m. Malinda Bumgardner, Randolph Carter, Mary Randolph Spotswood, Janet Carter, Brooke. The following is from one of the daily papers: "Maj. Randolph Carter Berkeley, of the U. S. Steamship Prairie, a native of Staunton, Va., the son of a Confederate officer, was the first American to land at Vera Cruz at the beginning of hostilities."

Admiral Fletcher in his report to Congress June 7, 1914, said, "I have similarly to record that the conduct of Major R. C. Berkeley U. S. M. C. (et. al.) were eminent and conspicuous in command of their battalions. These officers were in the fighting of both days and exhibited courage and skill in leading their men through the action."

(6). Emma Cornelia, b. April 8, 1849, d. i.

(7). Millard Filmore, b. Feb. 2, 1853, and d. in Augusta County March 6; 1914. He owned the old Gilkeson home "Hillside," and was for many years a prominent and useful citizen of Augusta County. An Elder in Bethel and later in the First Presbyterian church, Staunton, Va. He m. Betty Eskridge. Three children, viz., Eskridge (1914), of Bluefield, W. Va., m. Elizabeth Jones, of Staunton.

Jane, m. L. W. Wilson, a civil engineer and Maslin.

JOHN GILKESON of Bethel, who m. about 1820 Jane, was

THE GILKESON FAMILY.

probably a brother of William J. Gilkeson. They were dismissed from Bethel Oct. 16, 1830. Two sons, viz:

Isaac, b. March 4, 1824.

David, b. Sept. 13, 1827.

DAVID GILKESON, of Staunton, who m. Elizabeth Gilkeson, dau. of Hugh, may also have been a brother of William J.

III. JOHN GILKESON lived in Frederick County. He m. SARAH VANCE Dec. 1, 1793. Revolutionary soldier. Commissioned Major May 3, 1780. Buried at Opequon. See Cartmell's History of Frederick County. Nine children:

1. Elizabeth, b. Nov, 13, 1775, m. John White, of Winchester, Va.
2. Margaret, b. March 28, 1777, m. Reynolds, of Ky.
3. Mary, b. Aug. 8, 1779, m. a Limerall, of Ky.
4. Susan, b. Aug. 22, 1781, d. s.
5. Col. John Gilkeson, b. Sept. 15, 1783, and d. Feb. 27, 1856. M. Lucy Davis. Four children:
    (1). William D., m. a Baker of Winchester, Va. Four children, viz., William D., Jr., James, Frances and another son.
    (2). James. em. to Missouri.
    (3). Lucy m. a Woods, of Staunton. Dau. Janet Woods. A childhood playmate of Woodrow Wilson.
    (4). John, em. to St. Louis.
6. Sarah, b. March 20, 1785. m. William Gilkeson, her first cousin, son of Hugh. Four children.
7. Nancy, b. June 24, 1788, m. Stephen Davis.
8. Martha, b. Nov. 24, 1790, m. Rev. A. A. Shannon, of Kentucky.
9. James, b. Aug. 5, 1793, m. Bell. Three children:
    (1). John Bell Gilkeson, of Moorefield, W. Va.
    (2). J. Smith Gilkeson, of Winchester, Va., m. a Cabell. Two children:
        a. Mrs. A. M. Baker, Winchester, Va.
        b. Henry, em. West.
    (3). Robt. B. Gilkeson, Romney, W.Va. Two children:

a. Henry B. Gilkeson, Romney, W. Va.

b. Edward Gilkeson, Parkersburg, W. Va. Mar. a dau. of the late Rev. G. W. Finley, D. D. Dau. Margaret.

IV. SAMUEL, Revolution soldier; qualified Captain Aug. 4, 1779. Hugh Gilkeson, 1797, wrote from Frankfort, Ky., to his wife Elizabeth that he had been unable to find any trace of his brother Samuel.

V. DAVID, a Revolutionary soldier, died in prison.

VI. ISAAC, m. a Shanklin. Son John lived in Greenbrier County.

VII. EBENEZER, m. a Shanklin. Dau. Margaret. Lived Greenbrier County.

VIII. SUSAN, m. John Armour.

IX. MARTHA, m. Alexander Galt, of Pennsylvania. Two sons, John; and William, who had two sons, W. R. Galt, and A. W. Galt, of Pequea, Pa.

X. NANCY, m. William Vance.

XI. JANETTA, m. Thomas Marshall, of Virginia. Son, William, who m. his first cousin, Margaret Gilkeson, of Greenbrier.

FAMILY OF ROBERT GILKESON.

ROBERT GILKESON came to Augusta a generation ahead of the Bethel family. The two families are not known to be related. There is a tradition that they met for the first time on the boat en voyage to America. His deed for 400 acres of land on a branch of Middle river, near North Mountain, is dated Jan. 22, 1747.

He is mentioned in the records of Aug. 20, 1746; constable, 1756. He d. 1775. His will is recorded Staunton, Va. Chalkley III, p. 141. "Well stricken in years." Wife, REBECCA. Three children, viz:

I. ARCHIBALD, wife SARAH, from the Calfpasture. His will was proven 1782. Was private in Captain Patrick Martin's militia 1756. Surveyor of highways 1767. Five children, viz:

THE GILKESON FAMILY.

1. Robert; was living in Augusta 1799.
2. James; Revolutionary soldier at Battle of Point Pleasant, 1774, in Capt. Robert McClanahan's company. See Dunmore's Wars.
3. Hugh; was living, Augusta, 1791.
4. Rebecca, m. Oct. 25, 1797, George Moffett.
5. Francis, possibly the oldest son, b. Oct. 29, 1769, and m. Oct. 1, 1799, Mary, dau. of James Hogshead. She was b. May 19, 1777, and d. Sept. 6, 1860. Francis d. Jan. 21, 1842. Eight children, viz:
   (1). Hugh, b. Oct. 4, 1800, m. Matilda Hogshhead. Three children, David F., Thomas, Mary.
   (2). Grace, b. Sept. 21, 1802, m. Thos. C. Poage,
   (3). James, b. Feb. 14, 1805, m. Rebecca Trimble. Two children, William G., Margaret.
   (4). Jane B., b. Dec. 20, 1806, m. Silas H. Hogshead.
   (5). Malinda, b. Dec. 4, 1808, m. James McClung of Bethel church.
   (6). Robert G., b. July 9, 1811, m. Margaret Shields.
   (7). Rebecca A., b. Sept. 10, 1813, d. s.
   (8). Francis, b. 1816, m. Martha A. Crawford. Seven children,
   a. William F., Elder in Tinkling Spring; m. first, a Caldwell; son, Crawford, and several daughters; m. 2nd Lou Smith. Died 1912.
   b. Aurelius R. Elder Presbyterian church, Churchville, Virginia.
   c. Mary P.
   d. Nannie C.
   e. Emma A.
   f. Sallie.
   g. Carry S.
II. Margaret, m. David Hogshead.
III. Isabel, m. Hugh Brown.

THE HUMPHREYS FAMILY.

PHILIP HUMPHREYS was martyred Nov., 1558, at Berry St. Edmonds, Co. Suffolk, England, for denying the supremacy of the pope and rejecting the mass.

JOHN HUMPHREYS, sixth or seventh in descent from PHILIP, m. Margaret Carlisle, a cousin. They were both of Co. Armagh, Ireland. Eleven children, among them:

I. David Carlisle Humphreys (1741-1826) m. about 1770, Margaret Finley, d. of Wm. Finley and a niece of Rev. Samuel Finley, D. D., (1716-1766) born in Co. Armagh, President of College of New Jersey, now Princeton University. Ten children:

1. Margaret Finley, b. 1773, m. Nov. 24, 1797, Samuel Blackwood. Six children.

2. John, b. about 1775. Em. to Indiana.

3. Ann, b. about 1778, m. Oct. 26, 1798, Archibald Rhea. Died young.

4. Betsey, b. about 1780, m. Feb. 19, 1801, Sam'l McCutchan, Elder in North Mountain Church. Ten children.

5. Polly, b. about 1782, m. David Gilkeson.

6. Samuel, b. 1785, Elder in Bethel. Mar. Margaret, dau. of John Moore of Rockbridge. Nine children.

(1). Caroline, b. 1813, m. Robert Tate Wallace.

(2). Rev. James M. Humphreys, b. 1816. Dau., wife of Rev. Sam'l Gammon, D. D., missionary in Brazil.

(3). David Carlisle, 1817-1848.

(4). Capt. John Moore Humphreys, b. 1820. Co. I, 52 Va. Regiment, C. S. A.

(5). Samuel, b. 1821, died in Arkansas.

(6). Dr. William Humphreys, b. 1823, m. a dau. of Rev. Francis McFarland, D. D., pastor of Bethel. Son, Prof. D. C. Humphreys of W. & L. U. Dau. Theta, m. Rev. G. T. Storey.

(7). Howard A., b. 1826.

(8). Margaret Jane, b. 1829, m. 1851, Hon. Wm. Donald. D. June 28, 1914. Son, Sam'l Donald, of Staunton, Va.

THE HUMPHREYS FAMILY.

(9). Rebecca Weir, b. 1832, m. April, 1853, James Alexander McClure.

7. Tirzah, b. 1787, m. 1815, James S. Willson of Rockbridge, an Elder in Mount Carmel Church.

8. Hannah, b. 1789, d. s., buried at Old Providence.

9. Infant, died.

10. Aaron Finley, b. 1794, Elder in Bethel. Mar. Nancy Sterrett, (1803-1881) dau. of James Sterrett of Hebron, Augusta Co. Seven children, among them,

(1). Margaret Finley, b. Oct. 5, 1829, m. Geo. W. McClure.

(2). Isabella Sterrett, b. Jan. 17, 1831, m. John A. Gilkeson.

WILLIAM HENDERSON, gentleman, son of John Henderson, gentleman, of Fifeshire, Scotland, m. February 7, 1705, Margaret Bruce. Their third son, Samuel Henderson, b. Nov. 28, 1713, d. Jan. 19, 1782. For his will, see Chalkley III, p. 163. Jane, his wife d. 1800. Chalkley III. p. 219. Nine children, viz. Alexander, Andrew, David, Florance, James, Rebecca, Samuel, Jr., Sarah, William. Andrew, m. April 7, 1796, by Rev. John McCue, Margaret McClure. Em. to Blount Co., Tenn.

GEORGE HUTCHESON and his wife ELEANOR came from Lancaster Co., Pa., and settled on Long Meadow, joining Samuel Pilson. His deed is dated Feb. 21, 1738. One of the first Elders of Tinkling Spring, 1740. Died intestate, 1766. Issue:

I. John, b. about 1740, m. June 25, 1764. Son, John, Jr.

II. GEORGE, 2nd. b. about 1742, m. about 1770, Anne McClure. Issue:

1. John, m. March 9, 1793, Margaret Finley.

2. Eleanor, m, April 1, 1794, Joseph Henderson, Jr.

3. George, 3rd. m. December 4, 1798, Betty Stuart.

4. Margaret, b. 1785, m. 1807, her cousin Isaac Hutchinson of Greenbrier. See p. 25.

III. WILLIAM, settled in Greenbrier Co., sons, Isaac, (above) and George, who m. Jan. 7, 1786, Margaret Campbell.

IV. JAMES, whose dau. Jane, m. Nov. 26, 1801, Capt. Thos. Caldbreath.

THE STEELE FAMILY.

ANDREW Steele, d. testate 1764. Mentions four children, viz: ELIZABETH who possibly m. John McClure, p. 25, SARAH, ROBERT, SAMUEL, b. about 1724 and died testate, 1799. Seven children.

I. MARY, II. SALLY, III. Samuel, who d. June 8, 1837.
IV. JAMES, whose dau. Jane m. 1799, Robert Carden.
V. Catherine, who m. Thomas Jackson.
VI. Jenny, who m. 1787, Peter Alexander, p. 187.
VII. Robert, whose dau. Mary, m. 1801, Wm. McCormick.

DAVID STEELE, b. about 1700, d. testate 1747. Wife JANNET. Eight children, viz: JANNET, REBECCA, MARTHA who m. Teas, ISABELLA who m. about 1745, Moses McClure. SAMUEL d. testate 1796, five children.

I. MARY, m. a Rankin; son Samuel Steele Rankin.
II. JENNY m. Col. Cunningham; son, Samuel Steele Cunningham.
IV. Catherin, m. 1799, William Handly, Jr.
V. NANCY, m. 1799, Joseph Evans.
VI. WILLIAM.

ROBERT, died testate, 1800. Six children, MARY; ELEANOR who m. an Allen; MARTHA who m. Thomas Paxton; JOHN, father of John Jr., WILLIAM, DAVID b. about 1755, lived at Steele's Tavern. He m. MARY dau. of Samuel Steele; their dau. Jane m. 1796 George McCormick.

THOMAS, died testate, 1803. Six children, viz. Catherine, Jane, Rosanna, Sally, Robert and William.

NATHANIEL, d. testate 1796. Wife Rosanna. Six children viz: I. Eleanor who m. Capt, David McClure.
II. Rosanna who m. 1782, Samuel McClure. P. 136.
111. Mary, who m. Halbert McClure. P. 137.
IV. Daughter who m. Archibald Blackburn; dau. Rosanna.
V. MARTHA, who m. 1789, Robert Cooper, dau. Elizabeth.
VI. NATHANIEL, d. 1802. Son Nathaniel, 3rd, dau. Sally.

THE STEELE FAMILY.

SAMUEL STEELE, (1709-1790). m. a Fulton, aunt of Robert Fulton, the inventor. Seven children:

A. JAMES (1735-1802), m. Sarah Wright. Five children.

I. ANDREW, (1766-1832), m. June 18, 1795, Elizabeth Tate, dau. of Capt. James Tate.

II. SARAH, 1769-1827, d. s. Probably the Sarah Steele that reared John McClure. P. 49.

III. MARTHA, 1771-1855, m. Daniel Henderson. See p. 51.

IV. SAMUEL 1773-1835, m. Fanny Hunter. See p. 51.

V. JOHN, d. s. 1804. Chalkley III, 225.

B. SAMUEL, 1736-1808, m. Sarah Hunter. Dau. Catherine, m. 1787 John Thompson, of Rockbridge.

C. ANDREW, (1743-1800), wife Mary. Six children.

I. ANDREW, Jr., m. 1798, Martha Crawford.

II. JOHN, m, 1800, Polly Bush.

III. POLLY, m. 1800, Andrew McClure, Jr.

IV. James, V. Sarah, VI. Jean.

D. MARY, m. David Steele.

E. MARGARET, m. 1788 David Buchanan.

F. MARTHA. G. SARAH (1737.1808.)

THE TATE FAMILY.

MAGNUS TATE, the first of the name in Virginia, emigrated from the Orkney Islands, North of Scotland and landed at Philadelphia May 20, 1696, eventually locating in what is now Jefferson County, West Virginia. He is said to have died in September, 1747. Sarah Tate m. 1779 Bishop James Madison, of Rockingham County.

JOHN TATE, the founder of the family in Augusta County, settled near North Mountain October, 1744, and was "acquainted with the lands for two years before he came to live in the neighborhood." See Chalkley III, p. 31.

In addition to his farm, he owned and operated a mill. His name occurs very frequently in the Augusta records; as Justice in 1784, and Overseer of the Poor 1786.

His wife, MARY DOAK, belonged to the well known Augusta family. Brother-in-law of Francis Beaty.

He died March, 1801. Five sons and a daughter. Circuit Court Wills, Book 1, p. 41.

I. THOMAS TATE, b. about 1740, m. October, 1764, Jane Campbell, dau. of Charles and Margaret Campbell and a sister of Gen. Wm. Campbell. His son,

1. Charles Tate, m. his first cousin, Mary Tate, d. of Gen. Wm. Tate. Three sons,

(1). Charles Tate, 2nd, married and left,
    a. Charles Tate.
    b. John Tate, who m. Rebecca Tate, of Augusta County. Son, Friel Tate.
    c. Nannie, who m. Major David Graham, of Graham's Ford, Va. Eight children.
    d. Thomas Leonidas Tate, member Board of Visitors V. M. I., member Virginia Legislature and Senate; Ruling Elder. He m. Lucy Gilmer. Lives, Draper, Va.

(2). Leonidas Tate.

(3). Dr. Thomas M. Tate. Member Virginia Senate. Father of Thomas Green Tate, Culpeper, Va.

II. JAMES, b. about 1744; alumnus Augusta Academy.

THE TATE FAMILY.

Captain in Revolutionary War. Participated in the battle of Cowpens; killed in battle of Guilford March 15, 1781. His company was composed of men from Bethel and Tinkling Spring congregations. Schenck says: "Capt. Tate, of Virginia, so distinguished at Cowpens, received a ball which broke his thigh." He m. Sarah, dau. of Edward Hall. Five children:

1. John, b. 1774, and d. Missouri about 1866. He probably m. about 1800, Mary, dau. of William Anderson England. Grandfather of Rev. John C. Tate, Presbyterian minister, Clarksville, Tenn., and Rev. L. B. Tate, Korea.

2. Col. Isaac Tate, of Callaway County, Mo., m. Jane, dau. of Daniel and Martha (Steele) Henderson. Grandparents Mr. John N. McCue, Auxvasse, Mo.

3. Elizabeth, b. Oct. 28, 1778; m. June 10, 1795, Andrew Steele.

4. Polly.

5. Sally.

Sarah Tate, his widow, m. May 4, 1785, Hugh Fulton and em. to Flemingsburg, Ky.

III. ELEANOR, baptised by Rev. John Craig, Nov. 5, 1747. Mar. about 1770, Benjamin Stuart, b. 1736, son of Archibald and Janet Brown, sister of Rev. John Brown. Came to Augusta 1742. Five children:

1. Maj. Archibald Stuart of War of 1812. M. first, Polly Alexander, dau. of Francis Alexander and Elizabeth McClure. Three children, Andrew Alexander who m. Sarah McClure, Ellen who m. Jas. Brooks, parents of Charles Brooks and Mrs. Mary Booker; Martha died single. He m., second Mary Henderson; two children, Mary, wife of Sam'l Steele, and Benjamin, who m. Clem Willson. The latter family lives in Fort Worth, Texas.

2. John, em. to Indiana.

3. Nancy, m. an Alexander.

4. Mary, m. Nov. 5, 1788, John McClung.

5. Elizabeth, m. Dec. 4, 1798, Geo. Hutcheson, Jr. She m. second, Dr. James Allen. See Stuart Family, Waddell, p. 367.

THE TATE FAMILY.

IV. JOHN, bap. by Rev. John Craig, Feb. 26, 1749, Alumnus Augusta Academy; member Virginia Legislature, Trustee Staunton Academy 1792. Married about 1774, JANE. Died Dec. 1802. Nine children:

1. John, Jr., m. Feb. 27, 1794, Betsy McClanahan, dau. of Elijah McClanahan, Sr., and Lettice Breckenridge. Em. to Ky.

2. Isaac, em. to Kentucky after 1800.

3. Mary, m. Samuel Finley, Sep. 20, 1796.

4. Nancy, m. Adam McChesney, Jan. 10, 1800, dau. Jane Eliza.

5. Ellen, m. John Finley of N. Carolina.

6. Drucilla, m. Rev. John D. Ewing.

7. Jane, m. Jacob Van Lear, Augusta Co.

8. Elizabeth, m. Nov. 5, 1823, John Moffett.

9. Clorinda.

JANE, the widow lived in the Bethel congregation, near Greenville. Died June 1834.

V. GEN. WM. TATE. Physician. Educated at Augusta Academy and College of New Jersey. Revolutionary soldier. Mar. Nancy (Agnes) Mitchel of Phila. Three children:

1. Dr. Mitchel Tate.

2. Dorcas who m. John Campbell.

3. Mary who m. Charles Tate.

Thomas and William Tate emigrated from Augusta to Washington Co. Va. 1783.

VI. ROBERT, b. March 1753, and d. July 9, 1832. Farmer. Revolutionary soldier. Married about 1775 Margaret, (a famous beauty) dau. of John McClung and Elizabeth Alexander of Timber Ridge. She was born Oct. 1755 and d. Sept. 23, 1839. Ten children:

1. James, b. 1781, d. July 15, 1857. Farmer. He m. first, Rebecca Baxter, dau. of Capt. Geo. Baxter of the Revolution and Mary Love, dau. of Col. Ephraim and Elizabeth Love. Sister of Dr. Geo. Addison Baxter of Washington College and Union Theo. Seminary. Four children:

THE TATE FAMILY.

(1). George Baxter, b. 1809, m. Mary Young. He d. March 9, 1837. Dau. Mary George Baxter Tate.

(2). Robert, d. s.

(3). John Addison, b. Jan. 12, 1815, and d. Nov. 21, 1854. Farmer. Member Virginia Legislature. He m. Apr. 7, 1836 Margaret Randolph b. 1819 and d. Feb. 1907; dau. of JOHN RANDOLPH, b. in Charlotte Co., Va., Feb. 26, 1790, and d. in Middlebrook, Va., Oct. 11, 1861, and MARY JANE FRAZIER (1797-March 25, 1849) dau. of John Frazier (1750-July 11, 1832) and Mary Frazier (1761-Jan. 18, 1843).

John Randolph, a cousin of John Randolph of Roanoke, ran away from home at sixteen, settling in Augusta. He was a merchant and became wealthy. Letitia, a daugter m. Wm. F. Smith, of Greenville, parent of J. Ran., Mrs. Anna Lilly, Mrs. Wilson Brown, Mrs. Mary Randolph, Mrs. Lee Christian and Prof. W. Ballard Smith of McDonough, Md. His son, John T. Randolph, m. Anne Farish, lived and died Charlottesville, Va., leaving four sons, viz., William who m. his first cousin, Mary Smith, parents of Edmund and Bruce Randolph of Augusta Co., Thomas F., Dr. John, and Walter Randolph.

John A. Tate and Margaret Randolph left three children:
  a. Mary Jane, m. Dr. John M. Tate, of Greenville, Va.
  b. Rebecca Friel, b. Oct. 1, 1839, m. John Tate, of Wytheville, Va.
  c. Letitia Margaret, b. Feb. 19, 1844, m. John W. Gilkeson.

(4). Margaret Amanda, b. Jan. 12, 1815 (twin) and d. Nov. 6, 1836. Mar. July 9, 1835 Charles Lewis Peyton, an Elder in Bethel, son James Peyton of Greenbrier, m. an Eskridge.

Jas. Tate m. second, Mrs. Charlotte Beale. Two children:

(5). James Allen, b. July 14, 1832, d. i.

(6). Col. Wm. Poague Tate who m. first, Margaret, dau. of Joseph Kayser of Alleghany Co. Two children:
  a. Isabella, b. Sep. 6, 1843, m. Charles Cameron. Four

THE TATE FAMILY.

children, viz., Margaret m. John Opie, Jr., of Staunton; Ellen, m. Robt. Palmer; Charlie.

Mar. second, Sarah Christian, who m. second, Rev. W. T. Richardson, D. D., Editor Central Presbyterian; dau. Nellie Tate, now Mrs. Talbott, of Waynesboro, Va.

b. Margaret, b. May 27, 1846, m. Cyrus Creigh.

2. John, m. first, Nancy, only dau. of Wm. Moffett, of Augusta. Seven children:

(1). Major Wm. Moffett Tate, an Elder in Bethel and in Staunton. Mar. first, Mattie Frazier; 2nd, Kate, dau. of Dr. Addison Waddell. Son, Addison Waddell Tate.

(2). Robert, em. to Illinois.

(3). Dr. John Tate, of Greenville, Va.; mar. Mary Jane Tate. Eight children.

(4). Dr. James Tate, father of Miss Nannie Tate of Mary Baldwin Seminary.

(5). Melancthon. Em. to Florida.

(6). Margaret, m. Dr. Steele, Illinois.

(7). Elizabeth, m. Joseph Hite, Illinois.

(8). Rebecca, m. Blackburn, Illinois.

3. William, m. Elizabeth McClung. Son, William; em. to Florida.

4. Elizabeth, m. Sept. 5, 1793, Col. James Allen. Em. to Michigan.

5. Mary, b. 1777 and d. June 23, 1856; mar. April 24, 1794, Samuel Wallace, parents of Robt. Tate Wallace, et al.

6. Ellen, m. Samuel Patterson; d. Jan. 9, 1865.

7. Phœbe, m. Samuel Wilson, Brownsburg, Rockbridge County. Three children, viz: Esteline who m. Andrew McClung, parents of Jas. McClung, of Lexington, Va., and others. Sally who d. s.; and Rebecca who m. Col. Sterrett, of Rockbridge, parents of Mack and Tate Sterrett.

8. Rebecca, m. Reid Alexander, Rockbridge Co.

9. Isabella, born 1795, died Dec. 31, 1818; m. John B. Christian.

10. Sally, d. s.

THE WALLACE FAMILY.

This family is famous in the annals both of Scotland and Ireland. We find Rev. Jas. Wallace, pastor Presbyterian Church Urney, Co. Donegal 1654-74, and James Wallace, elder in Donaughmore, Co., Donegal, 1672-1700.

JAMES WALLACE, doubtless of the Donegal family, settled in Augusta county 1748; mar. Elizabeth daughter of John Campbell and Elizabeth Walker, He died 1780. It seems that he had but one son, viz

WILLIAM WALLACE, who m. JANE, dau. or John Hunter and who died intestate, 1779. Five children viz:

JAMES, WILLIAM, SAMUEL, FRANCES and MARY.

Samuel, m. May 24, 1794, by Rev. John Brown, Mary, dau. of Robert Tate and Margaret McClung. Six children,

1. Eleanor, m. June 13, 1839, Samuel Withrow.
2. Jane, m. Thos. Webb.
3. William, m. Mary Shields. Issue:

(1) John Samuel, b. Nov. 14, 1859, d. s.

(2) Francis Robert, b. Jan. 21, 1852; m. Dolly Shields. Three children, viz. Alexander, married and lives in Kentucky. William Allen, m. Mary Tate Gilkeson; Elizabeth.

(3) Elizabeth, m. James McFarland of Staunton, Va. Parents of Frank Patterson and Wallace McFarland.

4. Elizabeth, m. Jan. 27, 1824, Archibald McClung.
5. Mariah, m. Dec. 29, 1825, Benjamin McClung.
6. Robert Tate, m. Jan. 26, 1832, Caroline Humphreys. Five children.

(1) Margaret, b. March 11, 1833, m. Benjamin McClung

(2) Mary Tate, b. Sept. 27, 1834, m. John P. McClure. A grand-daughter Mary Margaret, dau. of John Marshall McClure and Mary Scott Storey, was b. July 21, 1914.

(3) Eleanor Amanda, b. 1837, d. s.

(4) Cornelia, b. Feb. 10, 1839, m. Jas. B. Smith. Parents of Ella, wife of Chas D. McClure, and Mrs. William F. Gilkeson, and others.

(5) Jas. William, b. June 20, 1844, m. 1st, Ophelia Willson. 2nd, Carrie Gilkeson. Three children by his first wife.

a. Clarence Willson, b. 1871, Chattanoogo, Tenn.

b. Dr. Harry M. Wallace, b. 1873. Greenville, Va., m. Lucy Baker of Staunton. Two children.

c. Robert Tate, b. 1881. Graduate W. and L. U., Student Union Theo. Seminary.

ADDENDA.

GILBERT MCCLURE, of Donoghmore, Co. Derry, Ireland, died teste, 1687.

RICHARD MCCLURE, of Gartan, Co. Donegal, Ireland, died testate, 1709.

ROBERT MCCLURE, gentleman, Co. Monaghan, Ireland, died testate, 1673.

DAVID MCCLURE, oldest son of David of Candia, N. H., m. a Miss Dinsmore and moved to Deering, N. H. Their son, David, 1758-1835, of Antrim, N. H., m. Martha Wilson. Eleven children. See p. 166.

DAVID MCCLURE, of Co. Donegal, settled about 1720 in Lancaster Co., Pa., where he died 1749. Wife, MARGARET. Three daughters and five sons. Of these, Alexander, David and John settled in Baltimore. The latter is probably the ancestor of John McClure of Baltimore, who m. Mary Anne Thornbury, whose daughter Georgianna Virginia m. John Thomas Schley, parents of Rear Admiral W. S. Schley. William, d. s., Randall m. ANN and had Alexander, James and John.

JAMES, JOHN and THOMAS MCCLURE settled in Ira, Vt., about 1779. Soldiers of the Revolution. James died in Middletown Feb. 22, 1815, aged sixty-seven. SAMUEL MCCLURE, soldier of the Revolution, lived in or near Newbury, Vt.

ROBERT MCCLURE (a brother William), b. 1734, settled in West Pennsboro, Columbia Co., Pa. He m. about 1755, Margaret Douglas. Three sons:

I. William, b. 1759, m. (1st) Agnes NcKeehan. Four sons. 1. John, who em. to Ohio. 2. Robert. 3. Alexander, who m. Isabella Anderson; parents of Col. Alexander Kelly McClure, b. 1828. He married 2nd —— McKeehan and had James, Samuel and Joseph.

4. William, who m. and had Robert, at one time curator of the United States mint at Philadelphia.

II. Alexander. III. Robert.

The Pa. Muster Rolls of 1776-1783 give:
Andrew McClure, private, 1778, Washington Co.
James McClure, private, 1781, Northumberland Co.
John McClure, private, 1781, Cumberland Co.
John McClure, 1778, Washington Co.
Samuel McClure, 1781, Northumberland Co.

New York McClures in the Revolution were:
Moses McClure, Continental line, 1st Regiment.
Moses McClure, Continental line, 2nd Regiment.
William McClure, Continental line, 2nd regiment.
William McClure, Continental line, Pawling's Regiment.

EARLY McCLURE MARRIAGES.

James McClure, Chester, Pa., m., 1763, Pattie Simpson, (1737--1807).

Dr. McClure, Lexington, Mo., m. about 1830, Eliza Hord.

Maggie McClure, of Mo., m. about 1870, Thos. Nathaniel Mudd, of the Maryland family.

Robert Arthur McClure, son of Dr. John E. McClure, of the Rockbridge family, m., February 18, 1846, Margaret Downey Morrison (1827-1854). Three children:

Belle Arthur, b. Dec. 16, 1846.
Anna May, b. June 24, 1849.
Charles Robert, b. August 10, 1852.

Lucy Eglantine McClure, daughter of James and granddaughter of Robert McClure of Ireland, m., March 10, 1840, Dr. James Hunter Merriweather, of Todd Co., Ky.

Mary McClure dau. of Henry McClure, and Mary Turner of Vt., m., March 28, 1855, Cyrenius Lyman.

Lucretia McClure (May 26, 1793,—Sept. 11, 1862) dau. of Thomas McClure, and Nancy Hunter of Bristol, Me., m., May 26, 1817, Edward Dyer Peters. See p. 163.

ADDENDA.

The following are Rockbridge Co., Va., marriages.

Agnes McClure m. Wm. Douglas Dec. 21, 1803. See p. 181.
Alex. McClure m. Betsy Paxton Nov. 17, 1808. See p. 140.
Arthur McClure m. Nancy Edmondson Jan. 5, 1798. See p. 142.
Catherine McClure m. Samuel McCorkle April 26, 1804. See p. 147.
Catherine McClure m. James Taylor Feb. 11, 1808. See p. 136.
David McClure m. Rhoda Jones Nov. 25, 1819. See p. 137.
Elizabeth McClure m. Jacob Morgan Jan. 1, 1801.
Fanny McClure m. Flamin Byers Jan. 26, 1804. See p. 147.
Isabella McClure m. Andrew Hall, May 20, 1799. See p. 136.
Jane McClure m. Joseph Paxton Nov. 22, 1792. See p. 147.
John McClure m. 1st, Jennet McClure June 2, 1808. See pp. 140 and 181.
John McClure m. 2nd, Nancy Cunningham Nov, 11, 1824. See p. 140.
John McClure m. Isabella Hall Nov. 28, 1799. See p. 136.
John McClure m. Ann McFall April 9, 1801; See p. 136.
John E. McClure m. Martha Parry Sept. 3, 1823. See p. 143.
Malcolm McClure m. Eliz. McClure Dec. 14, 1786. See p. 149.
Martha McClure m. John Jamison Feb. 27, 1800. See p. 136.
Mary McClure m. Nathan Dryden Aug. 30, 1785. See p. 181,
Mary McClure m. David Templeton May 10, 1791 See p. 141.
Moses McClure m. Eliz. Jones Feb. 18, 1812. See p. 137.
Nancy McClure m. Jas. H. Alexander April 13, 1820. See p. 147.
Nathan McClure m. Jane Porter Sept. 17, 1795. See p. 149.
Robert McClure m. Sophia Campbell Dec. 18, 1815. See p. 145
Sally McClure m. Wm. Grigsby Jan. 7, 1790. See p. 241.
Susanna McClure m. Jos. Stephenson Aug. 19, 1794. See p. 136.
Wm. McClure m. Sallie McClure Nov. 12, 1823. See p. 143.

On page 123, add,

VIII. Mary McClure, m. Dec. 27, 1785, Alexander McKinny. Em. to Kentucky.

IX. Eleanor McClure, m. Feb. 27, 1786, David Wilson. Em. to Kentucky.

ERRATA

Page 9, line 21, for "attained" read attainted.
" 25, line 24, omit "d. single 1837."
" 40, line 21, for "James White, Jr." read Charles White.
" 45, line 13, for "1829" read February, 1830.
" 60, line 24, for "1848" read 1840.
" 60, line 31, for "1772" read 1872.
" 65, line 25, for "Louisana" read Louisiana.
" 65, line 31, for "William" read William Frederick.
" 65, line 31, for "Frederick John" read John.
" 79, line 28, for "1907" read 1906.
" 90, line 25, for "1899" read 1799.
" 91, line 6, for "Satunton" read Staunton.
" 92, line 1, for "Samuel H." read Samuel.
" 103, line 4, for "Andred" read Andrew.
" 123, line 31, for "1872" read 1782.
" 135, line 24, for "childraen" read children.
" 140, line 22, for "ahd" read and.
" 142, line 8, for "father" read cousin.
" 142, line 10, omit " about 1775."
" 142, line 33, for "seven" read eight.
" 147, line 29, for "probably m. a Byars" read m. Jos. Paxton.
" 149, line 18, for "Malcoln" read Malcolm.
" 168, line 21, for "Bolly money" read Ballymoney.
" 186, line 10, for "carpulency" read corpulency.
" 189, line 7, for "Maky" read Mary.
" 192, line 37, for "Nichols" read Nicols.
" 194, line 9, for "Susan Jane" read Emeline.

INDEX.

INDEX.

Alexander Family, The, 184.
Alexander, A., 56.
Alexander, Andrew, 25, 186.
Alexander, Capt. Arch., 125, 186.
Alexander, Rev. Arch., 126, 187.
Alexander, Francis, 25, 188.
Alexander, James, 187.
Alexander, Rev. James, 184.
Alexander, Dr. Jas. H., 147, 228.
Alexander, Jos. McKnitt, 156.
Alexander, Robert, 15, 187.
Allen, Capt. Jas., 100, 149.
Allen, Mary, 149.
Allen, Rebecca, 100.
Allison, Robert, 136.
Applegate, Mary, 176.
Arbuckle, James, 25.
Arbuckle, John, 56.
Arthur, Helen W., 86.
Auld, Ann, 152.
Baker, Mrs. A. M., 208.
Baker, Mrs. N. J., 142, 181.
Barton, Rachel Sarah, 123.
Baxter Family, The, 188.
Bayliss, Allie, 77.
Beard, Esther, 187.
Beasley, Mrs. Stanley, 142.
Beaty, Francis, 215.
Beaty, John. 57.
Bell, Jane, 124.
Berkeley, Dr. Carter, 207.
Berryhill, Alexander, 202.
Berryhill, Wm. L., 55.
Blackburn, Dr. Grundy, 110.
Blake, Joan E., 165.
Bogurdus, Everardus, 193.
Bougere, Horace, 65.
Bowman, John, 117.
Bowman, S. McClure, 118.
Brown, Mrs. E. G., 116.
Brown, Samuel, 204.
Brubeck, Ada., 71.
Bumgardner Family, The, 190.
Burch, Jas. Harvey, 193.
Burlingham, Janetta, 131.
Callison Family, The, 72, 198.
Camfield, Ann F., 154.
Campbell, Elizabeth, 111.
Campbell, Sophia, 145.

Campbell, Gen. Wm., 215.
Capps, James, 55.
Carlisle, Hon. John G., 130.
Case, Warren, 65.
Christian, Lee, 194.
Clark, Fanchon, 88.
Collins, Ellen, 111.
Cotton, Mary, 194.
Coursey, Sam'l L., 28.
Craigmiles, Elizabeth, 167.
Crawford, Alexander, 149.
Crawford, Mary, 150.
Crockett, Margaret, 208.
Day, Anna E., 80.
Deekens, Dr. William, 207.
Dickerson, Frances, 103.
Doak Family, The, 84, 85, 126.
Draper Family, The, 208.
Draper, Mary, 208.
Dryden, David, 140.
Dubois, Alex. M., 115.
Dupuy, Nancy, 141.
Echols, Nora, 187.
Edmondson, Nancy, 142.
Elliot, Jean, 141.
Elliot, Martha, 138.
Ellison, Ophia, 192.
Eskridge, Elizabeth, 207.
Ewell, Henry C., 141.
Fauber, Barbara, 92
Finley, John, 19.
Finley, Rev. Samuel, 211.
Fish, Eunice K., 149.
Fishback, Mary, 110.
Fisher, Mrs. Margaret, 116.
Fleming, Maria Louisa, 121.
Fleming, Col. Wm., 121.
Frazer Family, The, 122.
Frazer, Edward, III, 122.
Frazier, John, 218.
Fulton, Betsy, 27, 195.
Fulton, James, 195.
Gammon, Rev. Sam'l, 211.
Gaston, Dr., 156.
Gaston, Elizabeth, 167.
Gaston, Mary, 156.
Gilbert, Chas. F., 117.
Gilbert, Edward A., 116.
Gilbert, Wm. W., 117.

INDEX.

Gilkeson Family, The, 204.
Gilmore, Rev. Robt. C., 148.
Gilman, Mary, 164.
Glendinning, Beulah, 114.
Graham, Maj. David, 215.
Grills, John, 208.
Guthrie, John, 204.
Hall, Andrew, 136.
Halstead Family, The, 192.
Hanna, William, 21.
Hardin, Eleanor, 208.
Harper, Amby, 129.
Harper, Jacob, 129.
Harris Family, The, 61.
Harris, Samuel, 156.
Harrison, Col. Benj., 130.
Hemphill, Paul, 155.
Henderson, Alexander, 27.
Henderson, Andrew, 26.
Henderson, Daniel, 27.
Henderson Family, The, 212.
Henderson, Mary, 139.
Hendrix, John M., 87.
Hendrix, Phoebe, 86.
Hicks, Mary, 114.
Hogshead, James, 220.
Holmes, Elizabeth, 124.
Houston, Robert, 156.
Hoyle, Mrs. Lucy, 132.
Humphreys Family, The, 211.
Hutcheson Family, The, 212.
Hutchinson Family, The, 25.
Hutchinson, Wm. T., 64.
Hyde, Thos., 206.
Ingles Family, The, 203.
Jackson, Thomas, 213.
Jans, Aneke, 193.
Johnson, Cynthia A., 88.
Jones, Elizabeth, 137.
Jones, Joseph H., 118.
Kerr Family, The, 124.
Kerr, Elizabeth, 90.
King, Adelle, 114.
King, Lucien, 114.
Kinkead, Edmund S., 123.
Kinkead, Elizabeth F., 123.
Knox, Mrs. Jno. B., 159.
Kurtz, Charles, 97.
Kyle, Beersheba Cobb, 36.
Lapsley, Rev. R. A., 79.
Larew, Lula Tate, 205.
Latimer, Mary J., 88.
Laughlin, S. H., 110.
Lawrence, Grove, 111.
Lawrence, Hal, 111.
Lawson, Mary P., 117.
Lewis, Duff J., 36.

Lewis, Eliza P., 86.
Lightner, Chas. T., 62.
Lightner, Frank B., 62.
Lightner, Geo. P., 201.
Logan, John, 130.
Logan, Col. John, 181.
Love, Col. Ephraim, 189.
Love, Mary, 189.
Lucas, Elizabeth, 88.
Lynn, Hugh, 204.
McCall, Elizabeth, 34.
McChesney, Jane, 144.
McClung, James, 219.
McClure, Aaron T., 179.
McClure, Abby, 178.
McClure, Abdiel, 174.
McClure, Abigail Caroline, 88.
McClure, Abigail Wheelock, 162.
McClure, Absolom K., 36.
McClure, Capt. Addison S., 182.
McClure, Adolphus B., 181.
McClure, Agnes, 18, 20, 135, 141, 181.
McClure, Alexander, 135, 136, 137, 140, 141, 164, 165, 178, 181, 221.
McClure, Rev. Alex. Doak, 158.
McClure, Alexander H., 188.
McClure, Col. Alex. K., 176, 222.
McClure, Alexander Stuart, 62.
McClure, Rev. Alfred Jas. Pollock, 178.
McClure, Alice Clara, 65.
McClure, Amanda, 110.
McClure, Amelia, 114.
McClure, Amos Harrison, 88.
McClure, Andrew, 4, 11, 18, 27, 89, 90, 110, 122, 132, 153, 174, 178, 181, 222.
McClure, Rev. Andrew, 98.
McClure, Andrew Fulton, 87.
McClure, Andrew Steele, 91.
McClure, Andrew Wellington, 65, 79.
McClure, Dr. Andrew W., 178.
McClure, Ann, 130, 178.
McClure, Anna C., 88.
McClure, Anna L., 179.
McClure, Anne, 25.
McClure, Anne Halstead, 78.
McClure, Archibald, 13, 167.
McClure, Arthur, 12, 14, 142, 178, 181.
McClure, Rev. Arthur, 141.
McClure, Asbury C., 148.
McClure, Benjamin, 174, 177.
McClure, Benjamin T., 78.
McClure, Bryan S., 182.
McClure, Carrie Pilson, 72.

INDEX. 229

McClure, Catherine, 136, 137, 175, 181.
McClure, Charles, 153, 175, 177, 178.
McClure, Col. Charles, 110.
McClure, Major Charles, 175.
McClure, Charles A., 169.
McClure, Charles C., 144.
McClure, Charles D., 72.
McClure, Charles E., 143.
McClure, Charles F., 165.
McClure, Charles F. W., 165.
McClure, Charles King, 114.
McClure, Charles P., 143.
McClure, Charles V., 169.
McClure, Lieut. Charles W., 113.
McClure, Clara Steele, 63.
McClure, Clay Pilson, 63.
McClure, C. P., 174.
McClure, Cochran, 152.
McClure, Cora T., 141.
McClure, Curtis, 146.
McClure, Cyrus W., 91
McClure, Daniel, 14, 148, 163.
McClure, Major Daniel, 182.
McClure, David, 4. 124, 160, 164, 166, 179, 221, 223.
McClure, Capt. David, 136.
McClure, Dr. David, 160.
McClure, Judge David, 176.
McClure, Rev. David, 161.
McClure, Dora Florence, 88.
McClure, Dorothy, 140.
McClure, Edmonia B., 115.
McClure, Rev. Edmund. 1.
McClure, Rev. Edward C., 1.
McClure, Edward Donald, 68.
McClure, Sir Edward S., 1.
McClure, Eleanor. 18, 21, 27, 92, 123, 177, 178.
McClure, Eleanor Wright, 117.
McClure, Elisha, 127.
McClure, Elizabeth, 25, 34, 90, 129, 131, 174.
McClure, Elizabeth F., 88.
McClure, Elizabeth Jane, 34, 110.
McClure, Elizabeth M., 45.
McClure, E. Mortimer, 151.
McClure, Esther, 20, 25, 127, 169.
McClure, Ethelyn Dell, 88.
McClure, Etta, 179.
McClure, Ewin, 3.
McClure, Felix, 115.
McClure, Finley, 128.
McClure, Finley Willson, 63.
McClure, Florence Adelle, 114.
McClure, Frances, 116, 176.
McClure, Francis, 153.

McClure, Capt. Francis, 151.
McClure, Rev. Francis, 12.
McClure, Francis Asbury, 88.
McClure, Francis Jasper, 150.
McClure, Frank Homer, 88.
McClure, Frank Wilson, 137.
McClure, George, 181.
McClure, George C. A., 151.
McClure, George D., 182.
McClure, George E., 91.
McClure, George Edgar, 63.
McClure, Gen. George M., 151.
McClure, Prof. George M., 150.
McClure, George W., 61.
McClure, Capt. George W., 181.
McClure, Gilbert, 8, 12, 221.
McClure, Halbert, 135, 137, 139, 141, 181.
McClure, Hannah, 140, 141, 181.
McCluer, Harry Scott, 137.
McClure, Henry, 148, 154.
McClure, H. E. Rev., 159.
McClure, Hepburn, 180.
McClure, Hettie Anne, 65.
McClure, Hosea Andrew, 88.
McClure, Hugh, 68, 123, 135, 157.
McCluer, Hugh Brock, 145.
McClure, Hugh S., 165.
McClure, Hugh Walker, 37.
McClure, Isaac, 123.
McClure, Isabel, 178.
McClure, Isabella, 13.
McClure, Isabelle, 143.
McClure, Jacob, 127.
McClure, James, 5, 9, 14, 17, 18, 20, 21, 25, 30, 90, 91, 126, 129, 135, 142, 157, 159, 166, 169, 170, 175, 177, 178, 181, 221, 222.
McClure, James A., 179. 183.
McClure, James Alex., 66.
McClure, Rev. James Alex., 79.
McClure, James Allen, 103, 111.
McClure, James Andrew, 38.
McClure, James Davis, 161.
McClure, James E., 182.
McClure, James Enos, 114.
McClure James Finley, 63.
McClure, Rev. James G. K., 166.
McClure, James Henry, 161.
McClure, James Madison, 140.
McClure, James P., 91.
McCluer, James Steele, 144.
McClure, Rev. James W., 188.
McClure, Jane, 79, 125, 174.
McClure, Jane Allen, 133.
McClure, Jane Ann, 62.
McClure, Jane Thompson, 78.

230 INDEX.

McClure, Jean, 18, 21, 25, 125.
McClure, Jean Weir, 68,
McClure, John, 4, 9, 12, 14, 21, 24, 49, 80, 91, 97, 129, 136, 141, 142, 149, 153, 154, 155, 169, 170, 174, 175, 178, 180, 181, 221, 222, 223.
McClure, Capt. John, 156.
McClure, Maj. John, 181. 182.
McClure, John B., 36, 176.
McCluer, John Cameron, 144.
McClure, Rev. J. Campbell, 2, 4.
McClure, Col. John D., 176, 182.
McClure, Dr. John E., 142.
McClure, John Finley, 63.
McClure, John F. L., 161
McClure, John Gilkeson, 79.
McCluer, Judge John G., 63.
McClure, John Howard, 68.
McClure, Rev. John J., 12.
McClure, John Marshall, 71.
McCluer, John Parry, 143.
McCluer, John Pilson, 69.
McClure, John Sandford, 88.
McCluer, John Steele, 144.
McCluer, Judge John Thos., 88.
McClure, Rev. J. T., 132.
McClure, Jonn W., 91, 138, 144.
McClure, Sir John W., 1.
McClure, John Wilfrid, 7.
McClure, Joseph, 132, 153, 156, 169, 181.
McClure, Joseph K., 36.
McClure, Joseph M., 170.
McClure, Josias, 134, 177.
McClure, Josie Charlton,
McClure, Katie, 65.
McClure, Katherine, 129, 178.
McClure, Katherine B., 79.
McClure, Lewis B., 77.
McClure, Lewis D., 80.
McClure, Lillie L., 63.
McCluer, Mrs. Lucretia 145.
McClure, Lucy Moore, 72.
McClure, Malcolm, 149, 181.
McClure, Malinda H., 65.
McClure, Margaret, 26, 88, 140, 178, 181,
McClure, Margaret Duff, 37.
McClure, Margaret Rice, 37.
McClure, Margaret R., 79.
McClure, Mary, 25, 40, 41, 120, 129, 141, 149, 164, 175, 178.
McClure, Mary A., 38.
McClure, Mary Alice, 68.
McClure, Mary Ann, 34, 88. 162.
McClure, Mary Fulton, 37.
McClure, Mary Lou, 62.

McClure, Mary Margaret, 220.
McClure, Mary Mildred, 72.
McClure, Mary Mitchel, 59.
McClure, Mary Stuart, 65.
McClure, Martha, 25, 40, 115, 136, 178.
McClure, Martha J., 143.
McClure, Martin, 4, 180.
McClure, Capt. Matthew, 155.
McClure, Mathew Thompson, 25, 28, 73, 80.
McClure, Mattie Lee, 37.
McClure, Michael, 128.
McClure, Michell, 4.
McClure, Milton, 114, 117.
McClure, Minnie M., 63.
McClure, Mitchel, 35.
McClure, Montgomery, 88.
McClure, Moses, 136, 137, 138, 140, 154, 181, 222, 223.
McClure, Nancy, 138, 163.
McClure, Nancy J., 38.
McClure, Napoleon B., 137.
McClure, Nathan, 135, 144, 182.
McClure, Nathaniel, 135, 139, 149, 166.
McClure, Nicholas J., 137.
McClure, Oliver S., 182.
McClure, Olivier, 180.
McClure, Oscar, 146,
McClure, Patrick, 133, 181.
McClure, Paul, 115.
McClure, Reba Belle, 68.
McClure, Richard, 12, 14, 153, 154, 175, 178.
McClure, Richard R., 150.
McClure, Robert, 4, 145, 176, 180.
McClure, Capt. Robert, 10.
McClure, Dr. Robert, 151.
McClure, Rev. Robert, 7.
McClure, Sir Robert, 1, 11.
McClure, Robert Alex., 147.
McClure, Robert Campbell, 143.
McClure, Robert G., 159, 182.
McClure, Robert Lewis, 87.
McClure, Robert M., 151.
McClure, Robert Shafer, 144.
McClure, Robert Vance, 71.
McClure, Robert W., 5.
McClure, Robert Wallace, 71.
McClure, Samuel, 11, 20, 92, 127, 136, 140, 149, 160, 167, 168, 177, 179, 181.
McClure, Capt. Samuel, 160.
McClure, Dr. Samuel, 182.
McClure, Rev. Samuel, 12, 151.
McClure, Samuel Campbell, 145.

INDEX. 231

McClure, Samuel Finley, 15, 68, 187.
McClure, Samuel Sidney, 167.
McClure, Sarah Alice, 36.
McClure, Sarah Barton, 123.
McClure, Sarah Jane, 88.
McClure, Sarah Johnson, 38.
McClure, Sarah Katherine, 80.
McClure, Sarah Steele, 65.
McClure, Sudie Louise, 114.
McClure, Susanna, 11, 136, 138.
McClure, Susan Louisa, 38.
McClure, Susannah Willys, 162.
McClure, Theodore, 110.
McClure, Thomas, 7, 14, 81, 130, 131, 132, 140, 154, 163, 178, 221.
McClure, Sir Thomas, 9, 13.
McClure, Thos. Bumgardner, 78.
McClure, Thomas Mero, 88.
McClure, Thomas Mitchel, 38, 88.
McClure, Rev. Uncas, 146.
McClure, Virginia, 38.
McClure, Virginia Wallace, 71.
McClure, Wallace Mitchel, 37.
McClure, William, 4, 5, 7, 14, 21, 133, 134, 140, 150, 151, 154, 168, 177, 178, 180, 181, 221, 222, 223.
McClure, Dr. William, 157.
McClure, William A., 148.
McClure, Dr. William B., 151.
McClure, William Bainbridge, 152.
McClure, William Bittenger, 144.
McClure, W. C., 110.
McClure, William C., 148.
McClure, William Harvey, 139.
McClure, William Kyle, 36, 37.
McClure, William Preston, 137.
McClure, William Thompson, 80.
McClure, William Warren, 80.
McCullough, L., 120.
McCullough, Thos., 26.
McCorkle, Malinda, 191.
McCorkle, Samuel, 147.
McCormick, George, 213.
McCormick, Wm., 213.
McCown Family, The, 78, 199.
McCoy, Elizabeth, 129.
McCue, John N., 45, 216.
McEwen, Sarah H., 34.
McGilvray, Sadie, 64.
McIntosh, Bettie, 155.
McKay, Hannah, 169.
McKee, Jane, 144.
McKee, Mary, 205.
McKragan, Ann, 154.
McLaughlin, Margaret, 148.
McLure, Judge J. J., 154.
McNutt, F. A. R., 35.

McSparran, Mrs. A. B., 170.
Mackelduff, Elizabeth, 170.
Mateer Family, The, 195.
Mateer, Billy, 43.
Mellersh, Claude M., 117.
Mellersh, Neale, 117.
Miller, Agnes, 195.
Miller, Sallie Phipps, 37.
Mitchel Family, The, 195.
Montgomery, James, 153.
Montgomery, Richard, 34.
Montgomery, Susan, 33.
Moody, Rev. Hiram, 132.
Morrison, Matthew, 153
Murphy, Dr. Alexander, 191.
Murphy, Dr. James, 191.
Myers, Rev. Chas. F., 79.
Nay, Jonathan, 164.
Nay, Sarah, 164.
Orwig, Rose, 114.
Owen, S. S., 87.
Parry, Martha, 143.
Parry, Mary, 143.
Parker, Emma F., 114.
Patten, Thomas, 165.
Patterson, Andrew, 204.
Peck, Eleanor, 92.
Peebles, Jesse, 115.
Pemrock, Mrs. Wm., 170.
Peters, Rev. J. P., 163.
Peters, W. R., 168.
Pilson Family, The, 27, 49, 58, 201.
Pomeroy, Rev. Benj., 162.
Ralston, Elizabeth, 7.
Randolph, John, 218.
Reed, Joseph, 140.
Reynolds, Richard C., 65.
Rogers, Martha, 88.
Scott, Nannie, 192.
Shafer, Elizabeth, 144.
Shields, Jane, 140.
Shields, Nancy Jane, 137.
Shreckhise, M. McClure, 92.
Simon, Judge J. T., 149.
Skidmore, Rebecca J., 129.
Smiley, John, 140.
Smith, Capt. Bird, 203.
Smith, Edward Lewis, 68.
Smith, Mayme. 68.
Sprong Family, The, 193.
Sproul Family, The, 193.
Steele Family, The. 211.
Stribling, Rev. C. R., 79.
Storey, Mary Scott, 72.
Strong, Elizabeth, 91.
Stuart, Andrew A., 64, 67.
Stuart, Judge Jas., 133.

Stuart, John Alex., 143.
Stuart, W. C., 149.
Sullivan, Andrew M., 121.
Sullivan, Capt. Wm., 120.
Sullivan, Jas. Wilson, 121.
Talbert, Robert, 202.
Tate Family, The, 215.
Tannehill, J. F., 192.
Thompson Family, The, 201.
Trigg, Gen. Abram, 203.
Trimble, Elizabeth, 90.
Trimble, Jean, 140.
Trimble, Mrs. H. H., 87.
Trimble, John, 90.
Trimble, Rev. W. W., 204.
Trotter, Isaac, 125.
Vance, Sarah, 208.
Varnum, Martha, 166.
Walker, Lolah, 116
Walker, Mae, 112.
Wallace Family, The, 220.

Waller, Rev. C. D., 199.
Watson, Rev. Samuel M., 147.
Webster, Clara 88.
Weir, Jane, 194.
Weir, John, 194.
West, J. P., 88.
Wetzell, Mary, 128.
White Family, The, 40.
White, Wm. McKim, 110.
Wilcox, William, 65.
Wiley, Lawrence, 111.
Wilkinson, Jean, 180.
Wilson, Mary, 163.
Wilson, Sallie, 137.
Wilson, Rev. Samuel, 175.
Willson, Matthew, 198.
Woods, Dr. Levi, 115.
Woods, Janet, 208.
Wright, Eleanor, 89.
Young, Jessie Peel, 121.

CPSIA information can be obtained
at www.ICGtesting.com
Printed in the USA
BVHW040841080422
633568BV00010B/193

9 781375 539203